HOW TO GROW A ROBOT

HOW TO GROW A ROBOT

DEVELOPING HUMAN-FRIENDLY, SOCIAL AI

MARK H. LEE

THE MIT PRESS   CAMBRIDGE, MASSACHUSETTS   LONDON, ENGLAND

© 2020 Massachusetts Institute of Technology

All rights reserved. No part of this book may be reproduced in any form by any electronic or mechanical means (including photocopying, recording, or information storage and retrieval) without permission in writing from the publisher.

This book was set in Stone Serif by Westchester Publishing Services. Printed and bound in the United States of America.

Library of Congress Cataloging-in-Publication Data

Names: Lee, Mark H., author.
Title: How to grow a robot : developing human-friendly, social AI / Mark Lee.
Description: Cambridge : The MIT Press, 2020. | Includes bibliographical references and index.
Identifiers: LCCN 2019025740 | ISBN 9780262043731 (hardcover) | ISBN 9780262357852 (ebook)
Subjects: LCSH: Robots—Control systems. | Robots—Social aspects. | Machine learning. | Artificial intelligence—Forecasting. | Human-computer interaction.
Classification: LCC TJ211.35 .L43 2020 | DDC 006.301—dc23
LC record available at https://lccn.loc.gov/2019025740

10 9 8 7 6 5 4 3 2 1

CONTENTS

PREFACE   xiii

    Acknowledgments   xv

I   WHAT'S WRONG WITH ARTIFICIAL INTELLIGENCE?   1

  1   THE NATURE OF THE PROBLEM   3

    Acting and Thinking   4
    The Social Robot   6
    The Role of Artificial Intelligence   7
    Intelligence in General   7
    Brains Need Bodies   9
    The Structure and Theme of This Book   9
    Coping with the Pace of Change   16
    A Note on Jargon   18

  2   COMMERCIAL ROBOTS   19

    Domestic Robots and Service Robots   20
    Field Robotics   22
    Robotic Road Vehicles   23
    Medical Robots   26
    Swarm Robotics   27
    Entertainment Robots   29
    Companion Robots   30
    Humanlike Robots?   31
    Observations   32

3 FROM RESEARCH BENCH TO MARKET 35

Bin Picking 38
Biorobotics 40
Care and Assistive Robots 40
Affective Computing 41
Humanoid Robots 42
Why Has Industrial Robotics Been So Successful? 46
The Current State of Robotics 50
Observations 53

4 A TALE OF BRUTE FORCE 55

Searching through the Options 56
The World Chess Champion Is a Computer—So What? 58
Computer "Thinking" 62
The Outcome 63
Observations 65

5 KNOWLEDGE VERSUS POWER 67

How Can Knowledge Be Stored for Utilization? 70
Common Sense Knowledge 72
Search Is a Standard Technique 74
Symbols and Numbers 75
Learning to Improve 75
Feature Engineering 77
Observations 78

6 A LITTLE VISION AND A MAJOR BREAKTHROUGH 81

The End of Feature Engineering 86
What Happened? 91
Observations 92

7 THE RISE OF THE LEARNING MACHINES 95

The Evolution of Machine Learning 96
Data Mining in Supermarkets 97
Learning Algorithms That Learn Algorithms 100
Discovering Patterns 101
Big Data 102
Statistics Is Important, but Misunderstood 104
The Revolution Continues—with Deep Zero 105
Observations 109

8 DEEP THOUGHT AND OTHER ORACLES 111

AI Is a Highly Focused Business 112
Task-Based AI 113

Machine Oracles   114
Knowledge Engineering   118
Social Conversation   121
Observations   124

9   BUILDING GIANT BRAINS   125

Brain-Building Projects   126
Whole Brain Emulation (WBE)   128
The Brain Is a Machine—So What?   130
Basic Artificial Neural Networks (ANNs)   133
Different Approaches: AI and Brain Science   134
More Advanced Networks   137
Predictive Coding and Autoencoders   138
Issues with ANNs   139
Simulation Problems for Robots   141
Observations   143

10   BOLTING IT ALL TOGETHER   145

The Complexity of Modular Interactions   146
How Can Computers Represent What They Know and Experience?   149
The Limitations of Task-Based AI   151
General AI   151
Master Algorithms   152
Biological Comparisons   154
Superintelligence (SI)   155
Integrating Deep Artificial Neural Networks (ANNs)   158
Observations for Part I   160

II   ROBOTS THAT GROW AND DEVELOP   167

11   GROUNDWORK—SYNTHESIS, GROUNDING, AND AUTHENTICITY   169

The Classical Cybernetics Movement   171
Modern Cybernetics   174
Symbol Grounding   176
The New Robotics   177
Observations   179

12   THE DEVELOPMENTAL APPROACH—GROW YOUR OWN ROBOT   181

The Role of Ontogeny: Growing Robots   184
Sequences, Stages, and Timelines   185
Constraints on Development   188

Start Small and Start Early 191
The Importance of Anatomy 193
The Amazing Complexity of the Human Body 195
Autonomy and Motivation 197
Play—Exploration and Discovery without Goals 198
An Architecture for Growth 201
Observations 206

13 DEVELOPMENTAL GROWTH IN THE iCUB HUMANOID ROBOT 207

iCub—A Humanoid Robot for Research 208
Managing the Constraints of Immaturity 210
Vision, Gazing, and Fixations 211
Motor and Visual Spaces 213
Object Perception 215
Experiment 1—Longitudinal Development 215
Experiment 2—The Generation of Play Behavior 217
How Does It Work? 221

III WHERE DO WE GO FROM HERE? 229

14 HOW DEVELOPMENTAL ROBOTS WILL DEVELOP 231

How Developmental Robots Behave 232
Taught, not Programmed 237
Knowing Oneself and Other Agents 239
Self-Awareness Is Common in Animals 241
Robot Selves 242
Consciousness 244
Communication 246
Developmental Characteristics 247
Will All This Happen? 248
We Must Get Out More … 250
Observations 251

15 HOW AI AND AI-ROBOTS ARE DEVELOPING 253

Task-Based AI 253
Human-Level AI (HLAI) 255
Deep AI 257
Robot Developments 259
Social Robots 260
Artificial Human Intelligence (AHI) 262
Observations 264

16 UNDERSTANDING FUTURE TECHNOLOGY 267

    Rapid Growth—It's Not Really Exponential   268
    Growth Patterns in the Twenty-First Century—So Far   270
    Artificial General Intelligence (AGI)   272
    Deep Networks, Learning, and Autonomous Learning   273
    Are There Any Dead Certainties?   274
    Trust, Validation, and Safety   278
    The Product-Centred Viewpoint   279
    The Crucial Role of Humans   284
    The Ethical Viewpoint   286
    Lessons from Opaque and Unregulated Markets   288
    Observations   290

17 FUTUROLOGY AND SCIENCE FICTION 293

    Are We Smart Enough to Know How Smart Animals Are?   294
    What Kind of World Do We Live In?   295
    Futurology, Expert Opinion, and Meta-opinions   296
    Threats on the Horizon?   300
    Superintelligence and the Singularity   300
    Transhumanism—Downloading the Brain   302
    Imminent Threats   303
    Toward Dystopia?   306
    It's Not All Doom and Gloom!   309
    Threats in Perspective   310
    Final Remarks   310

APPENDIX: PRINCIPLES FOR THE DEVELOPMENTAL APPROACH   313
NOTES   319
BIBLIOGRAPHY   339
INDEX   357

To Elizabeth, with love and gratitude for living with this unwelcome interloper for so long.

PREFACE

I have always been fascinated by the relationship between machines and humans. In my youth, I was very interested in electrical machines and engineering generally, which, with hindsight, I now see as an incredibly creative discipline. But during my training as an engineer, I was particularly attracted to systems that appeared to mimic some aspects of human behavior. Even simple feedback systems can be captivating (e.g., watching a ship constantly correct its course while being buffeted about by winds and tides). This leads to questions like: What kind of mental machinery do humans actually use? I somehow managed to work on projects that combined engineering with human-centered issues: speech encoding, color vision processing, and autonomous control. I found that psychology was the vital missing element, and my PhD, in modeling sensorimotor control and coordination, was a precursor of the work described here. The relationships among computers, brains, and machines are many and fascinating.

When I tried to pursue this line of research, in the 1980s, it was completely out of favor. The new commercialism of artificial intelligence (AI) had just begun, and AI software was becoming highly marketable. Before then, the early pioneers of AI saw no large distinctions between human and computer intelligence and thought both could be studied together. The drive for software products and applications caused AI to become largely separate from the study of human cognition. I remember discussing my work with Andy Barto, Richard Sutton, and others at the University of Massachusetts–Amherst when they were in the middle of developing reinforcement learning. In the UK, it was very difficult to get any funding for basic, curiosity-driven AI research, so I started work on AI in industrial robotics, tactile sensing, and error recovery and diagnostic systems, still exploring how humans behave during such tasks and how AI could be used in robotics. Fortunately, the twenty-first century has seen a swing back to topics like learning, autonomous agents, and renewed attention to human performance, and there is now a strong global research community in the new field of developmental robotics.

I have written this book partly in response to some common misconceptions about robots and AI. A great deal of misinformation about robots and AI has been bandied about in the media, some of which is clearly unreasoned nonsense. While there will be some amazing technology produced in the near future, there are limitations inherent in AI, and there are ethical issues that involve us all.

However, this is not a negative story because, after explaining the difficulties and nature of the problem, I then go on to describe an alternative approach that is currently delivering results and showing real potential for developing general-purpose robots. I give details from my own and related research to show how psychology and developmental ideas can get nearer to humanlike behavior—which is fundamentally general purpose but can adapt in order to learn to accomplish specialized tasks.

I try to avoid the minutiae of the latest technology or subtle changes of fashion; instead, I concentrate on the trends, the way technology develops, and the role of technology in our lives. I feel that it is important that everyone should be much more engaged in appreciating the technological developments of the day in order to get a better perspective on what is feasible, what is reasonable, and what is wild hyperbole. By applying basic common sense to many of the claims and predictions, we can evaluate, and therefore have a much larger influence on, the role of technology in our lives (and future lives). This book is a response to these concerns and contains some of my findings and insights into these fascinating problems. I hope that it will give a perspective that offers a better way of approaching some of the wider issues raised by intelligent robots.

PREFACE

ACKNOWLEDGMENTS

There are many colleagues, friends, and students that have helped me in my career, and it is impracticable to list them all. This includes colleagues in the Computer Science Department at Aberystwyth University, as well as many others in universities and other organizations in the UK, Europe, Japan, and the US. Research is a truly global community. I am grateful to all these friends and colleagues who generously supported, influenced, and encouraged the approach described here. I should particularly thank those who were directly involved in the work reported in this book: Raphael Braud, Fei Chao, Kevin Earland, Tao Geng, Alexandros Giagkos, Richard Gunstone, Martin Hülse, Ian and Tom Izzett, Suresh Kumar, James Law, Daniel Lewkowicz, Sebastian McBride, Qinggang Meng, Marcos Rodrigues, Patricia Shaw, Michael Sheldon, Qiang Shen, James Wilson, and Xing Zhang. Of course, the opinions expressed here are solely my responsibility, as are the errors and omissions.

Others who have helped in more mysterious ways include Anna Borghi, Merideth Gattis, Frank Guerin, Kevin Gurney, David Llewelyn, Giorgio Metta, David Middleton, Kevin O'Regan, Peter Redgrave, Peter Tallack, Raymond Tallis, and my late but dear friends Brendan McGonigle and Ulrich Nehmzow. Forgive me for not listing all the others; you are not forgotten, and I am grateful to you all.

I much appreciate the expertise and kind support of Marie L. Lee at MIT Press, and I thank all the copy-editing and other staff who have helped in this project.

The research described here was mainly carried out during four projects: two funded by the British Engineering and Physical Science Research Council, and two funded by the FP7 program by the European Commission (EC). The iCub humanoid robot originated in an EC-funded project, and the EC had the foresight to support the supply of these robots to new research projects, with the result that iCubs are now being used in more than twenty-five robotics research laboratories worldwide.

I am grateful to the Royal Society of Arts for permission to reuse and modify material taken from my article "A Frame of Mind," *Royal Society of the Arts Journal*, no. 2, 2018, 46–47. This appears in chapter 9, in the section entitled "The Brain Is a Machine—So What?"

I am grateful for the use of many libraries, particularly six local libraries. This includes the Bronglais Postgraduate Center and the wonderful resources of Llyfrgell Genedlaethol Cymru, the National Library of Wales. Finally, I am most grateful for having such a positive, helpful, and brilliant family, and particularly for my dear wife, Elizabeth, who has encouraged and supported this project in so many ways.

I

WHAT'S WRONG WITH ARTIFICIAL INTELLIGENCE?

1

THE NATURE OF THE PROBLEM

Common sense—Ordinary sensible understanding; one's basic intelligence which allows for plain understanding and without which good decisions or judgments cannot be made.
—*Wiktionary*, July 16, 2019

Keepon is a tiny robot, only 12 centimeters high. It's made of two balls of yellow silicone rubber: a head and a body. All Keepon can do is roll its head from side to side, nod backward and forward, and rock and bounce its torso. But humans find Keepon very interesting and get involved in "conversations," as Keepon nods its head and jigs about in excited response to their vocalizations.

Keepon has no arms or legs. It cannot move around or handle objects. It has no facial muscles; its face is simply two tiny eyes and a circular nose (actually, the eyes contain small cameras, and the nose is a microphone). Any expression or communication has to be through bodily movement or gesture.[1] Yet this extremely minimal robot is quite engaging for humans. Why?

This phenomenon of engagement is also seen in animated films, where moving line drawings, minimal representations, somehow convince us that the depicted characters actually exist. The answer, of course, is that they are animate; they move in exactly the right way—the way that living

1.1 Keepon displaying eye contact and joint attention (looking at an object that the human has just looked at). Used with permission from Professor Hideki Kozima (Tohoku University).

creatures move. Walt Disney created an industry, and made a fortune, from this effect.

Keepon is captivating because it *looks at humans when they talk* and also performs what's known as *joint attention*—it looks at the same objects that the human looks at, giving the impression of shared interest (see figure 1.1). It's not necessary for a social robot to pretend to be human, or even resemble us. We are happy talking to all kinds of animals: Just look at the fuss we make over a puppy or a kitten! But we do need that animal-like behavior and response.

This raises a big question: If Keepon and some other robots can give a passable imitation of animal movements, why do the vast majority of robots seem so clunky, so "mechanical"? If we want future robots to become useful assistants, companions, or supernumeraries, then they will have to be much more humanlike, friendly, and engaging.

ACTING AND THINKING

Robotic devices have fascinated both scientists and the public at large for a very long time. Some of the earliest ones were treated as religious marvels, such as the automaton monk of 1560, a small figurine that prayed while "walking" across a table top.[2] From the seventeenth century onward, automata became very popular as entertainments, and the eighteenth century was a golden age for autonomous mechanical figures and animals. Thanks to the skills of watchmakers, really impressive models were constructed, such as the life-size Silver Swan of 1773, which moves its head and neck realistically, preens itself, and catches a wriggling fish from the simulated moving water on which it sits.[3] The automatic human figures of those times often played musical instruments or danced; sometimes they wrote or drew on paper; but they always mimicked human behavior. The key to their attraction was that their actions were humanlike.

From the clockwork toys and automata that so amused our ancestors to the amazing science fiction creations seen in modern film and television, we have always had a deep empathy for machines that can mimic human actions. This fascination extends across a wide range of movement styles, from total mimicry to imitations of bodily actions. We enjoy watching the grace and elegance of the human form in athletics or ballet, but we are also intrigued by the motions of factory machines that assemble components or paint cars ("poetry in motion," as I saw emblazoned on the side of a heavy lifting crane).

It's not just movement: We are also fascinated by displays of purely mental activity. Champion chess players, quiz masters, and expert players at word and number games all draw large audiences; and thinkers like Isaac Newton, Charles Darwin, and Albert Einstein are revered as scientific heroes. Future robots will have to be smart too—very smart. If they are going to keep up with humans, then they will have to understand us and what we want, and this requires a lot of brain power. They will have to perceive us, as humans, and continually learn about our changing wants and needs. And all this must happen in real time; that is, *in time with us*— not too fast or too slow. We soon ditch technology when it gets boring, inflexible, or unresponsive.

These two requirements, *appropriate behavior* and *mental ability*, are the subject of this book. Some robots are good at moving; some computers are good at thinking; but both skills are necessary for really intelligent robots. We need robots that humans can engage with, robots that are perceptive, animated, and responsive to us. This book explores if and how this can be achieved; and what is necessary to support such an advance. In short, the question is: *How can we build intelligent robots that think and behave like humans?*

THE SOCIAL ROBOT

Among the robots that already exist, very few even attempt to behave like humans. Keepon is in a very small minority. Most robots can be called *autonomous agents*, but very few are *social agents*. This is the issue that this book examines: Robots that think and behave like humans must be social agents, freely interacting with humans and accepted by humans.

Just to be clear, a social robot is more than just a fancy interface, avatar, or conversational system. Social robots act like a companion or friend, with frequent interactions over an extended period. They won't necessarily look like, or pretend to be human, but they will understand human gestures, behaviors, and communications and be able to respond in ways that we find socially satisfactory. They will have limited "emotional" behavior but they will learn and adapt through their social life, recognizing different people and remembering their personalities from previous encounters. Of course, they will also do useful work, but rather than being programmed for specific tasks, they will be trained by their users and learn through demonstration and human guidance, as we do. If we want to create long-term, useful robot assistants, then social interaction is the key.

We will see in chapter 8 how competence in certain kinds of dialogue has been demonstrated. But this is a narrow application: The big question is how far we can progress toward sustained and meaningful social intercourse.

This book is not concerned with any specific application of social robotics; the social robot is proposed here as a milestone for the highest level of robotic achievement: thinking and behaving like humans. This is a research goal, not an application target. For example, I don't advocate humanoid robots for the care of elderly people; this important area still requires the human touch.

But once we know how to create social robots, once we have developed the technology, there will be plenty of scope for applications, practically anywhere that human-robot interactions regularly occur.

THE ROLE OF ARTIFICIAL INTELLIGENCE

So, how can we build social robots? What techniques are required, and what knowledge do we need? The growth of artificial intelligence (AI) is widely seen as the answer. AI is viewed as the bedrock on which future robots will be created. (I use a simple definition of the term here: *AI* refers to computer systems that achieve results that are assumed to require intelligence.)

It's true that AI is making enormous strides, producing impressive advances and products. For example, IBM's Watson system can answer questions posed in everyday language. Watson dissects the question and assembles answers using huge data sources: over 200 million pages of data, including the full text of Wikipedia! This quiz machine, which won $1 million on the *Jeopardy!* game show playing against two highly expert contestants, is now being adapted for use in health care, finance, telecommunications, and biotechnology.

In other areas, like speech and image analysis, new deep learning techniques have advanced AI. Computer recognition of images now demonstrates error rates of less than 1 percent. Image processing, like analyzing X-ray images or labeling the content of pictures, is producing impressive applications and products. This new boom in AI is generating a great deal of investment. Small AI companies are frequently bought for hundreds of millions of dollars. The heads of big companies believe that they need an AI department; and often go out and buy one—or several! Robotics has always involved AI, and many future robot systems will be powered by these new AI methods.

INTELLIGENCE IN GENERAL

There are three broad strands to human intelligence: the intrinsic abilities and adaptive potential of the individual brain, the joint results of combining several brains in social interactions, and the collective cultural knowledge produced by and available to humanity (see box 1.1). AI has mostly concentrated on building versions of mental skills and competences found in the brain: an *individual* brain. This ignores the interactions *between* brains that form a crucial context for human behavior

Box 1.1
Sources of intelligence

> **Brains** The origins and expression of intelligence are found in animal brains. This obvious fact has led to most AI systems being constructed as a single agent, thus reinforcing the idea of the individual as the sole source of intelligence.
>
> **Social interactions** Any activity that requires a group of individuals to complete, such as problem-solving in emergency situations, will involve social cooperation and organization. Social conversation involves communication, shared understanding of events and experiences, and learning through language.
>
> **Culture** The term *culture*, in this context, refers to the cumulative collection of all written text, data, diagrams, reference materials, and other works that humanity has compiled. Through reuse and training, such knowledge enables individuals to become more intelligent and also, through study, to discover further knowledge.

(box 1.1). In our early years, our comprehension, worldly understanding, knowledge, skills, and other cognitive abilities all depended on the interactions of at least one attendant caregiver. Without parental care, we do not develop and thrive and often do not even survive. In adult life, we form groups, societies, organizations, and networks, which are important for our continued development. Experts consult their peers and predecessors, thereby increasing their intellectual skill and knowledge. The third aspect of intelligence is culture. By *culture*, I do not mean trends, fashions, or popular activities, but rather the totality of recorded human experience across the global population. Language gives humans these extensions of intelligence into social and cultural dimensions; this is perhaps our only real difference from other animals, as will be discussed in later chapters.

Many AI systems are individual packages of intellectual skills, such as game-playing programs. On the other hand, speech and textual processing systems (e.g., translation) rely on enormous data banks of externalized knowledge. Cumulative cultural knowledge is recorded and disseminated in all kinds of ways; recently, this has become important for AI and is often known as *Big Data*—large data repositories that capture vast quantities of text, images, and other information.

So brains and culture (or intellect and knowledge) are well represented in AI, but social interaction between individuals is less in evidence as a research area. Social systems *are* studied in AI, two examples being the pioneering work on the robot Kismet (Breazeal, 2002) and the research program of Dautenhahn (e.g., Dautenhahn, 2007) (see Fong et al., 2003 for a general review). Social interaction is a feature of human interfaces, interagent communications, and robot-human interactions, but these nearly all focus on specific tasks or applications; they often do not consider the social sciences and the way that humans actually behave and think. AI generally has to face the fact that human intelligence is not entirely contained within the skull; it is also diffuse and distributed, and it often requires social interaction and close cooperation.

BRAINS NEED BODIES

Many scientists and philosophers have argued that bodies are essential for intelligent agents, and modern robotics has contributed evidence to this concept. Embodiment, together with constructivism and primacy of action, are key principles in understanding how the life process, the life cycle of individuals, affects their cognitive growth. We will examine these ideas in Part II.

Social agents, both humans and robots, need real-time, real-life experience: They have to have a body! The body is not just an input-output device; rather, it is the source of the experiences that shape cognitive processes. You have to *possess* a body *and* be prepared to talk about it. Social robots need a model of the "self"; they have to know that they have a body and be aware of its properties and behavior. Then they can take a subjective stance; they can view their experience from their own perspective, not that of their designer. This may appear to be a radical viewpoint. It requires us to examine what AI offers, look at alternative approaches, and consider and compare the possible outcomes.

THE STRUCTURE AND THEME OF THIS BOOK

In this book, I have avoided the two extremes of (1) the detailed and arcane complexities of modern technology and (2) the heavy philosophizing that AI attracts. Instead, I have adopted an *engineering* approach that focuses on the purpose of the products and how they serve our interests. I believe most of these issues can be explained in practical terms. I aim to ask simple but probing questions to cut through the politics, bias, and exaggeration and expose the most fundamental issues upon which judgments can be made. The results are given in a section entitled "Observations" at the end of each chapter.

It is necessary to understand the basic ideas before we can form an opinion about them, and unfortunately, the complexities of computing technology often seem to obscure the concepts and principles at the heart of the matter. A general understanding, with sufficient relevant information, allows common sense to draw sensible and sound conclusions. For example, car drivers need to know (and actually are mandated by law to know) quite a lot about how a vehicle behaves in response to its controls. There are various things to learn about how and when to apply the brakes on bends, at speed, and in wet conditions. Drivers need to master complex relationships between the controls and the braking effects desired. These are often subtle and take time to learn. But drivers do not need to know what kind of brakes they are using, how they operate, or what engineering design considerations they embody. That information isn't necessary to be a good driver or to make informed opinions about what features are desirable for our transport needs.

The same should be true with computing technology. We may know that there is some kind of digital processor inside the latest device, but we don't need the fine details; we need a good idea of what it does, a rough idea of how it does it, and an understanding of how we can use it—what use it is to us.

Thinking and behaving are characteristics that don't often occur together in machines. At the "behaving but no thinking" end of the spectrum, we find robotic devices that can perform all manner of physical movements but are controlled, by either fixed recordings (e.g., paint-spraying robots) or remote human operators (e.g., nuclear reactor maintenance robots). In both these cases, the robot is not considered to be thinking. At the other extreme, we find computers that just answer questions (like HAL in *2001: A Space Odyssey*, or IBM's Watson, described in detail in chapter 8). These are technically not robots because they don't have bodies and don't move about.

We must consider both robotic hardware and pure thinking machines because, as we will see later, behaving and thinking are intimately interrelated in humans and animals, and probably need to be so in robots too. You have to be able to think *and* move the pieces to play a game like chess;[4] but who would have thought that physically moving the pieces would turn out to be the harder task to automate?

PART I: THE LIMITATIONS OF AI FOR ROBOTICS

The first part of this book looks into the state of current robotics and AI. We begin in the next chapter, with a brief overview of a range of commercial robot offerings that is presented in order for readers to appreciate the scope of modern robotic devices and what they do. This gives a picture of their current capabilities, achievements, and limitations. This is followed in chapter 3 by an overview of research robotics, including the way that innovation is supported and funded. A summary table shows two key factors that determine the difficulty of implementing successful robots: the nature of the tasks that they are set to do, and the complexity and chaos in the environments in which they have to work.

Because AI is key to future robotics, part I contains several chapters looking at how AI actually works, what it can deliver, and what it has trouble doing. Each chapter introduces a key topic that readers can use to build an understanding of the ideas and issues in modern AI and their relevance for robotics.

It turns out that there are some serious obstacles when it comes to humanlike behavior. Despite media claims, AI is nothing like human intelligence and has not made much progress in moving in this direction. AI is very powerful for specific tasks and excellent for solving individual problems, but it still struggles to reproduce the general-purpose intelligence characteristic of the human species. No program has yet been written that can pretend to be a human in any convincing way for more than a few minutes. The AI chapters lead to a striking conclusion: *Engineering methods and AI are not sufficient (or even appropriate) for creating humanlike robots.*

Humans are *generalists*, unlike many animals. We readily adapt to fit different niches; we even adapt niches to suit us. This requires a wide-ranging ability to adapt to new situations. Humans are autonomous, motivated by internal drives as much as external forces, and the key to our flexibility is our open-ended learning ability. This allows us to face new challenges, learn about them, and find ways of dealing with them. Some animals are prewired by their genes; ducklings will follow the first "parent" they see after hatching. But humans are able to learn how to master completely new tasks from scratch. Even our failures teach us important lessons. So a humanlike robot really requires general-purpose intelligence that has the potential to deal with any task.

General-purpose intelligence is the "holy grail" of AI, but thus far it has proved elusive. Despite sixty years of research and theorizing, general-purpose AI has not made any progress worth mentioning, and this could even turn out to be an impossible problem (at least, impossible for software-based AI). However, even if general-purpose AI could somehow be realized in a computer (which I doubt), this would not be enough to solve the problem of thinking and behaving like humans. This is because human intelligence is *embodied*: All our thoughts, sensations, intentions, relationships, concepts, and meanings only make sense in terms of a body. Our brains do not run abstract codes that control our behavior—that's the wrong metaphor. Brains (i.e., our thoughts) are deeply dependent on, and shaped by, our bodily experience, including social and cultural experience. We are so fully integrated into our living environments that our internal thinking structures mirror and match these in an almost symbiotic way.

The AI chapters build up from basic concepts toward recent advances in Big Data and Deep Learning (chapter 7). These developments have made some remarkable progress and are being heralded as the solution to all remaining AI problems. The idea is that AI learning methods are now (or soon will be) powerful enough to learn any task required, and thus evolve into general-purpose intelligence. This would be a serious and important event in the history of AI, and so it has profound implications. Toward the end of part I, chapters 9 and 10, on brain models and AI integration, respectively, explain the issues with general AI and the nature of agency and embodiment. These two chapters contain the core of the argument against AI in social robots.

It is possible to skip some of the AI details, but do be sure to read the list of short points ("Observations") at the end of each chapter. These state the bottom lines that summarize all the important things to remember, and they cumulatively build up the book's arguments.

PART II: DEVELOPMENTAL ROBOTICS

The obvious question then is: If AI faces obstacles in building humanlike robots, are there other ways to go? Are there any other paradigms we can explore? The second part of this book presents a very different approach. This new viewpoint says: If you want a humanlike robot, why not start from what we know about humans? The idea here is not to "build your own robot," but rather to "grow your own robot." Don't load up your robot with piles of preprepared human knowledge; allow it to explore for itself and build its own knowledge about the things that come to have meaning for it, as we do. This is an embodied, enactive approach that allows cognitive functions to grow in step with experience. The idea is to shift the source of knowledge and meaning from the robot designer, or from Big Data, to the robot itself. Inspiration and data for this approach come from studies of early infant behavior, where many new and striking ideas on learning and interactive behavior can be found.

Infants have very limited perceptual and motor skills at birth but rapidly become physically and socially adept. They learn to communicate, they build an understanding of themselves as a "self," and they learn to recognize other social agents. By modeling some of these mental growth processes, robots will be able to *develop* and learn social interaction abilities, rather than being given social algorithms by their designers or extracting behavioral patterns from human data files.

This part of the book is supported by my own research into development as a robot learning mechanism. Experiments on the iCub humanoid robot have refined the approach and produced interesting results. The robot has developed from very poor levels of sensory-motor competence (just a few reflex actions and weak sensory resolution) to gaining mastery of the muscles in its limbs, eyes, head, and torso, perceiving and reaching for interesting stimuli, and playing with objects. These longitudinal experiments have taken the robot from equivalent infant stages of newborn to about nine months. At present, the iCub can reach and play with objects, associate single word sounds with familiar objects or colors, make meaningful gestures, and produce new actions for new situations.

Fortunately, the way that babies play and interact is well documented in the large body of theoretical and experimental work found in developmental psychology. This provides both inspiration and fiducial milestones for the new research field of developmental robotics. The origins of this kind of robotics are explained in chapter 11, as is embodiment, which is fundamental and different from AI. Then, the way that the iCub works is described and the principles made clear. It becomes obvious that robots can have a kind of subjective experience because everything learned is based on active experience *as seen by the robot*. By developing learning, they generalize and refine their most basic experiences so that meaningful patterns and correlations are stored, while unreliable phenomena are discarded. Note that these robots develop individually; they may begin identical, but they will produce somewhat different mental models and behaviors. This reminds us of the uniqueness of humans: Each person has her or his own history of lifetime experiences, and these contextualize particular social interactions.

This "growing" methodology, involving repeated refinements through experimentation, is synthetic rather than analytic and, I believe, gives us a much better route toward human-friendly robots with more humanlike characteristics.

PART III: FURTHER AHEAD, LIKELY DEVELOPMENTS, AND FUTUROLOGY

The implications for the future of both AI robots and humanlike robots are outlined in the final part of this book. The likely near-term advances in behavior and performance are estimated and compared. For instance, AI robots are given goals by their designer; they have an objective stance because they are structured from without. So the contrast with developmental robots is sharp and revealing. For example, future AI robots will have vast data resources available to them. They will be able to answer all kinds of questions, they will solve problems, and they will learn. So entering an unknown room, recognizing objects, and carrying out tasks using the objects in the room will become quite routine. Their abilities with objects and environments will initially be much superior to developmental robots. But their behavior will not really be humanlike. They will use Big Data on human behavior to simulate common human patterns, often convincingly. But they will solve problems differently, reason differently, and construct different concepts and percepts. And these differences will be detectable, and actually *noticeable*, particularly in social situations. AI robots will know humans only as objects—active agents, but objects nonetheless. They will fail to empathize, to appreciate another social agent as the same as themselves. This is because models of self cannot be designed, but rather are dynamic and must be created from experience.

In contrast, developmental robots will have an internal model of "self" (their body space, their actions and effects, and their environment), and they will be able to match their experience of another similar agent or similar behavior with their own self-model and recognize a human as a separate agent: another "self." This allows much better interaction and social relationships because the robot can anticipate and appreciate why humans behave as they do. Learning will improve the model of the interactive partner, and exchanges will become more grounded in the context and understanding of both individual and joint intentions and goals.

After exploring these differences, I then take a somewhat wider look at the impacts that we might expect from this kind of technology. The inevitable spread of AI will affect all realms of human life. Such so-called disruptive technology offers both threats and benefits. I comment on some of the claims and alarms generated by the current research climate. This includes the threat to jobs, the role of trust, and the positive benefits of keeping humans involved in combined human/computer systems. It becomes clear that the threats from robotics and AI are part of the context of wider threats from general digital technology. In this part of the book, the observations spread out to apply to our total computing infrastructure.

The final chapters consider a number of projections for future progress and examine their feasibility. All sorts of nonsense is written in the media about robots, and it can be hard to know what to believe. Journalists love robotics stories because they allow huge scope for speculation, outrageous predictions, and scaremongering. Headlines like "Robots Could Be Deemed as an 'Invasive Species' and Threaten the Future of the Human Race Because They Are Evolving So Quickly," which appeared about a BBC program on July 25, 2017, are not uncommon. The news media continually produce stories about the dangers of intelligent robots taking over practically everything, and this is often backed up by futurists and academics who predict superpowerful breakthroughs at regular intervals. While amazing advances both in the power and scope of applications offered by new technology, do occur regularly, there is no reason to blindly project this success into areas that have repeatedly failed to live up to predictions. The fact that a great many technological problems have been solved in a very short time is not sufficient evidence that we are on the verge of solving *all* outstanding problems.

I try to examine the current situation from a less hysterical perspective. From an analysis of the documented predictions of over 200 experts, I categorize and average their claims to produce a metasummary. This reveals a more balanced view of future developments from practicing engineers who are working in the field and actually solving difficult AI and robotics problems. This includes a small group of the leading experts at the major high-tech companies—those with great experience, those who have made major breakthroughs, or both. They recognize the limitations of technological progress and offer a mature, positive, and more stable image of the future. Nevertheless, the threats are real, if somewhat different from the media hype, and we must all be involved in the safe and beneficial management of future digital technology.

COPING WITH THE PACE OF CHANGE

Three factors are responsible for the recent surge in AI applications: (1) the doubling of raw computer power every two years has been enhanced by special hardware for graphics and neural networks, as well as new hardware for running new data-learning methods; (2) massive data storage facilities are becoming common, providing huge, readily available data banks on all kinds of topics; and (3) programming and software tools are much improved, and it is getting easier to rapidly implement complex software for new applications.

THE NATURE OF THE PROBLEM   17

Box 1.2
Terminology for programs

> **Algorithms** are plans for a computation. An algorithm gives a simplified description of what a program does or should do to achieve a defined job. Algorithms are the language of software, used among designers, programmers, and developers. Algorithms are to programs as recipes are to cooking.[5]
>
> **Programs** are the written instructions for a computer that are intended to perform a specific task. Programs usually implement an algorithm. A wide range of programming languages exist for different uses (e.g., business, scientific, gaming, embedded, interactive, web-based).
>
> **Code** is a set of instructions running inside a computer. Programs are converted into code during the implementation of software.
>
> **Software** is a program or collection of programs that make up an application or system. *Software* is a mass noun, like *money*, so it frequently refers to many programs.
>
> **App** is short for *application software*. Apps perform a specific task and can be easily installed into a system. They are widely used for smartphone applications, but there are also web apps and general computer apps.
>
> **Bots** or **softbots** are programs that operate autonomously to find or record particular data on the internet or within a large data store.
>
> **Chatbots** are bots that interact with humans, often on the internet, in a conversational mode. They are used to assist customers on websites and commercial services. Personal assistant devices are another example of chatbots.

These advances mean that new ideas can be tried out more quickly, and previously impractical ideas become feasible. The range of tools and ideas available for creating new applications has increased enormously.

For these reasons, it is not sensible to write about only the latest developments; basing analysis on the very latest products is a futile effort. By the time this book is published, the very latest technology will be quite different. But what we can do is examine the growth trends and general principles behind what's going on. By looking back at past milestones, we can see how obstacles have been overcome and get general insight into how things develop. The history of a technology offers a perspective that is valuable for predicting and assessing the future. The more data that are available for consideration, the more confidence that we can have in any projections.[6]

A NOTE ON JARGON

I aim to avoid jargon, but I've already used the word *algorithm*! I will try to explain any overly technical terms that come up in the discussion. Actually, *algorithm* is a good example of a technical term that now has entered the vernacular. Once we talked about computer programs, then software, and now algorithms. At a colloquial level, all these can be taken as roughly equivalent, but to add a bit of discrimination, box 1.2 gives some details on the particulars of each term. Software usually consists of a collection of programs, and programs implement one or more algorithms; an algorithm is the essence of a computation. New terms have been coined for particular classes of application, as the last three illustrate.

I also try hard to avoid formulas and unexpanded acronyms—those dread, dead symbols that separate the technological laity from those "in the know."

For those who wish to follow up on some of the many fascinating topics, I include references to the relevant literature. I usually recommend either the classic papers or (for new concepts) introductory material, and other citations provide links to key technical sources.

2
COMMERCIAL ROBOTS

The *Encyclopaedia Galactica* defines a robot as a mechanical apparatus designed to do the work of a man. The marketing division of the Sirius Cybernetics Corporation defines a robot as: "Your Plastic Pal Who's Fun to Be With."
—Douglas Adams, *The Hitchhiker's Guide to the Galaxy*, 1978

In 2012, something big happened. A team of artifical intelligence (AI) researchers made a major breakthrough: They discovered how computers could learn for themselves what features to use for automatically recognizing images. They smashed through a very big barrier that had hampered AI right from the beginning. AI systems could now learn tasks without handcrafting by their designers. This general area, using new learning methods and masses of data, has become known as *Deep Learning*.

Deep Learning generated much enthusiasm for AI and also raised expectations, especially in the media. We are currently in a boom period for AI and robotics; companies are heavily investing in these technologies, and research laboratories are creating a new wave of exciting applications and products.

This means that robots are becoming more useful, with new and better abilities, and becoming much more involved in everyday life. Unfortunately, the sheer number of innovations makes it harder to assess the current situation and evaluate the roles and value of robot systems.

We need a framework for identifying and understanding the fundamental issues among the wild diversity of forms and functions in this creative field. We need to concentrate on questions like: What do robots actually do? What are their limitations? What have they got in common? What classes of problem can they solve? Understanding these issues will give us a better sense of the way things are going and what to expect in the future.

The key feature of any robotic application is the nature of the *task* that the robot has to accomplish. In addition, there are *constraints* that have been either (1) avoided to simplify the task or (2) applied to control or restrict the working environment. For example, an agricultural robot is much easier to design if it needs only to spray weed killer over all the crops rather than to locate and pull out each individual weed (as some robots now do). This constraint makes the task easier (but not for organic farmers). Alternatively, an environmental constraint might be applied by structuring the field so that the weeds are easily identified (possibly by organizing or marking the crops). Constraints help to simplify things. The real world is full of chaos, noise, and uncertainty, and any constraints that a designer can exploit will significantly simplify things; using constraints effectively is a major talent of skilled engineers.

Let's start by looking at the design and behavior of a selection of commercial robots and assess their basic abilities. In each case, think about the tasks that they perform (what do they actually do?) and any constraints that they exploit. The answers will emerge by the end of chapter 3.

We should first define a robot as an autonomous physical device, of any shape or form, that can sense and perform actions in its working environment. Note that the requirement for autonomy rules out remote-controlled so-called robots such as drones.[1]

DOMESTIC ROBOTS AND SERVICE ROBOTS

Two tasks now performed by commercial robots are lawn mowing and floor cleaning. A Scandinavian company, Husqvarna, was the first to market a robot lawn mower, and it is now a major player with several models that vary according to the size of lawn;[2] as an example, see figure 2.1.

These small, battery-driven mobile devices can cut several thousand square meters of grass, entirely without human involvement. Batteries

COMMERCIAL ROBOTS

2.1 A robot lawn mower, courtesy of Husqvarna.

provide less power than conventional mowers, and so they make many complete passes over the lawn, each giving a small trim, rather than one large cut. But this turns into an advantage because the smaller clippings don't need to be collected and are actually better for the lawn if left on the surface, similar to mulch. (It has been said that a really high-quality lawn is best cut with scissors, which is effectively what these robots do.) The mowers are quiet and run unattended; they determine their own cutting pattern, avoid objects, stay within a set region determined by a wire boundary marker, and return to their recharging station when they sense a low battery level. They can also use the Global Positioning System (GPS) to know where they are and can communicate with their owner via a smartphone. Robot lawn mowers are now standard products, offered in many garden suppliers' catalogs. They are made by a range of companies in France, Italy, Germany, the UK, and the US.[3]

Indoor floor-cleaning robots are also widely available and operate in a very similar manner. Instead of cutters, they use vacuum and brushes (for carpets) or wet mops (for hard floors). They cover areas bounded by walls, avoid falling down steps, and self-recharge. They are less expensive than mowers, and two of the main manufacturers, in Germany (Bosch) and the US (iRobot), are now well established in the global consumer market. This new market area has seen companies rush to bring many similar products to the market.[4] In the UK, six different brands were available in 2017, with the expensive (but very effective) Dyson 360 being the most highly rated.[5] Bigger robot cleaners are also available for large or public premises.[6]

These mowers and cleaners are both examples of *domestic robots*. Their specifications are very similar, involving the movement of a cutting or cleaning tool over a fairly well defined surface area.

They don't need legs (wheels work just fine), and the power and environmental requirements are not arduous. This means the task is well constrained (i.e., with very limited external interference) and not high in difficulty.

Another example based on the motorized mobile platform idea is the robot security guard. These robots wander around empty buildings or offices, like a human guard, and use various kinds of detectors, such as vision, ultrasonics, and infrared, to sense any humans or animals that might be present. They can then alert security staff through various networks and also may take video images of intruders. Again, a range of companies market this type of robot.[7]

A kind of role reversal of the security robot is the museum guide. These robots also work in public spaces, but they seek out people in order to give advice or assistance about the museum or venue. They have considerable entertainment value and tend to be developed ad hoc for particular places, and so they are not always commercial, but often rather experimental.

These indoor mobile robots are examples of *service robots*, which perform functions like the collection and delivery of office mail, medicines and supplies in hospitals, or robot servers in restaurants and couriers that deliver food. Most service robots have to interact with humans in some way and therefore need some social skills. This has usually been provided through fairly basic interfaces. Service robots have to deal with more complex environments than floor cleaners and so face somewhat more difficult tasks, while security robots face more complexity from dealing with unexpected human behavior.

FIELD ROBOTICS

Some of the most demanding robotic application areas are military and policing tasks, especially search and rescue and bomb disposal. The popular PackBot is a small, twin-tracked vehicle that can enter disaster areas, survive drops of 1.8 meters onto concrete, and right itself if it lands upside down.[8] Over 3,000 PackBots have been sold, and they are regularly used to examine suspicious objects, for search-and-rescue operations in earthquake zones, and inspection roles during nuclear cleanup, such as after the Fukushima nuclear disaster in 2011.

There are many other examples of such highly robust robots for operation in real-life natural environments; this area is known as *field robotics*.[9] The outdoor world is the realm of field robotics, covering all kinds of autonomous devices: on-road and off-road vehicles, flying machines (aircraft and balloons), and boats and submarines. Commercial successes have been achieved in areas such as mining and quarrying, surveying and environmental assessment, agriculture and horticulture, and search and rescue. For example, agricultural robotics is a rapidly developing area; there are now various kinds of robot systems that can monitor the growth of crops, weed between the rows, and harvest from the ground or from bushes or trees. A particular success in farming has been robot milking parlors, which allow cows to choose for themselves when they want to be milked, thus relieving dairy staff from working very long, unhealthy hours. Further, the cows are less stressed and give higher yields. Before this century, robot milking was considered to be either science fiction or wildly expensive, and yet now it accounts for one in three new milking systems sold to farmers. Clever engineering will continue to reduce initial costs and increase market adoption.

ROBOTIC ROAD VEHICLES

Self-driving cars are now a commercial proposition, although research versions proved the concept many years ago. In 1995, two experimental vehicles (a delivery van and a Mercedes-Benz saloon, both stuffed with computers) undertook long-distance field trials through live traffic on public roads. One drove 2,850 miles from coast to coast in the US (Pomerleau, 1997) and the other, in Germany, drove more than 1,000 miles from Munich to Copenhagen and back (Dickmanns, 2007). In both vehicles, the steering was visually controlled from cameras that sensed the road-marking patterns, and the Mercedes also had computer control of the throttle and brakes. Safety drivers and experimenters were in these vehicles all the time, but they rarely needed to take over for safety reasons.

All the major automotive manufacturers now have their own versions of self-driving cars, and they will soon become very common on the road.

A big stimulus for robot vehicles has been the US Defense Advanced Research Projects Agency (DARPA) challenge series. In 2004, driverless cars had to drive 150 miles across a desert track. No team's car reached the finish line, but the second competition in 2005 was completed by 22 entries. The third challenge, in 2007, was based on a mock-up urban course in an old air base. More recently, DARPA has organized other robotics challenges. From 2012 to 2015, a $3.5 million challenge (http://archive.darpa.mil/roboticschallenge/overview.html) was focused on recovery and rescue robots that could do "complex tasks in dangerous, degraded, human-engineered environments." Unfortunately, this was overambitious, and entrants did not rise to the levels seen in the other challenges.

Daimler Trucks scored a world first when, in 2015, it obtained a license to operate fully autonomous trucks on public roads.[10] This development can be seen as a step on an evolutionary road map that started with the now-familiar cruise control systems that automatically maintain constant speed on long stretches of clear, open road. We now have brake lights that get brighter the harder you brake, autodimming headlights, and tailgates that open by waving your leg under the rear bumper. We can now speak to our cars, and we can make gestures to control the temperature, entertainment system, or sunroof. (I wonder how "gestures" to other road users will be interpreted by the vehicles, and vice versa!)[11] Systems such as lane departure warning, radar-based collision avoidance, and night recognition of pedestrians are also available. The next step is handing over to a robot the remaining driving controls: steering and acceleration.

Daimler, the parent company for Mercedes, has seen its technology mature over many years on a variety of vehicles. Google entered the self-drive arena in 2009,[12] and a race is now on between the car industry and the technology sector to capture the new self-drive markets. Initially, Google's cars were lightweight and slow—so slow (with a top speed of 25 miles per hour) that they got cautions from the police. But Google is not going to be left behind: In December 2016, they spun off all their autonomous car development work to a new company: Waymo. Recently,

Daimler joined Audi and BMW in a deal to acquire Nokia's software-mapping company. This will be necessary to compete with Google's powerful mapping and geographic information systems.

Apart from known intense activity at the big car manufacturers, General Motors, Ford, Nissan, Renault, Toyota, Audi, BMW, Hyundai, Volvo, and Peugeot, there are smaller firms and start-ups that look very promising, such as Tesla Motors, nuTonomy, and Mobileye (recently bought by Intel). The emerging maturity of electric cars and hybrids, as well as the temptation for other high-tech firms like Apple to get involved, should accelerate development.

It is important to distinguish self-driving vehicles from completely driverless vehicles. There is a five-level taxonomy of autonomous vehicles (level 0 means no automation), ranging from level 1 (very minor assistance) to level 5 (full driving autonomy).[13] Most cars under development are at level 3 or 4, which requires a driver to be present, pay attention, and take control if needed. Vehicles with *no one on board* (level 5) are another matter entirely; the safety, legal, and ethical issues involved are much harder to solve.

Just assume for a moment that all the technical problems have been overcome; imagine sending your own vehicle to the supermarket to collect your order. Even though your personal robot car might be safe with regard to road and traffic behavior (although this is difficult enough), events might occur for which the system would be completely unprepared. Examples include taking verbal instructions from a farmer to avoid an accident, getting stuck in mud in the farmer's field, handling gridlock, dealing with vandalism, or just avoiding erratic drivers. Such events are almost certain to occur, but designers and manufacturers cannot consider *all* possible scenarios. This is known as a *long-tailed data problem*, wherein most of the data is clustered around a set of common cases, but there is a "long tail" of rare cases that are serious if they do happen.[14] The problem is that individually the long-tail cases are *extremely* rare on an individual basis, but in the aggregate, there are so many of them that there is a good chance of one occurring.

This might not matter in a constrained application area, such as a university campus, a factory complex, or a shopping mall, where simple procedures such as emergency shutdown and remote communication could be sufficient, but releasing unmanned robot vehicles "in the wild" (into the world at large) is sure to cause problems. In such situations, human drivers use their common sense and the background knowledge they have. AI research is working on this, but the problem now has nothing to do with driving and everything to do with general intelligence. Despite the manufacturers' confidence in ongoing research (http://www.daimler.com/innovation/en/) and their aim to commercialize fully autonomous cars so that "the car comes to the driver," it may take considerable time before these technical, ethical, and legal issues are sorted out to everyone's satisfaction.

MEDICAL ROBOTS

Medicine is another burgeoning application area for robotics, and although the equipment is expensive and tends to be very specialized, patients are now routinely offered robotic procedures for a range of treatments. For example, the da Vinci surgical robot provides minimally invasive surgery for prostate, bladder, and other internal organ operations (www.davincisurgery.com/). Other robotic devices assist surgeons in orthopedics and bone implants. Of course, these robots are not fully autonomous and surgeons are always in control, but they can provide more accuracy and better results than humans, especially where very high precision is required. For example, when fitting metal bone implants, the surface of the bone needs to be shaped to match exactly the implant that is to be glued onto it. The machining of the bone can be driven from computer data on the shape of the implant and reproduce the exact surface contours of the implant, thus producing a much better fit than human skill can achieve.

There are many other related applications, often known as *health-care robotics*, including hospital support robots and rehabilitation robots. But medical robotics is a complicated area. While some results are excellent, others involving soft tissue are not yet perfected, and the very high cost of medical equipment is a major concern.

SWARM ROBOTICS

Robot swarms are collections of many robots, often identical, that have to cooperate to perform tasks that are beyond the physical abilities of a single person or robot.

Robot soccer is a classic example of swarm robotics. The RoboCup competition (www.robocup.org/) has been held as an international tournament since 1997; it runs a series of leagues for different robots of different constructions and abilities, ranging from small LEGO robots, wheeled mobiles and larger designs, up to humanoids that, instead of just pushing the ball, actually kick it (!). This tournament series has been very successful in promoting and publicizing robotics and encouraged many schoolchildren and other amateurs to build their own robots.

In the early years, the robots had trouble finding the ball and not falling over, but progress has been impressive—although the stated target of holding a match between robots and humans by 2050 is an interesting open question! There are now many such contests involving robot designing and building, and they have given excellent publicity to the field.

Other forms of swarms include insectlike mini-robots, formation-flying aerial vehicles, and sea- and submersible-based swarms. As the numbers in a swarm increase, the demands on processing power in each agent decrease. Ants and termites build large structures and perform multiple functions, but they clearly have limited individual nervous systems.

An interesting case of an extreme robot swarm is the Kilobot swarm, developed at Harvard University (see figure 2.2). Kilobots are tiny circular robots, only 33 mm in diameter, which move in a kind of shuffle by vibrating their spindly wire legs. But they have an onboard processor, memory, and battery, and they can communicate with nearby Kilobots using infrared, as well as by sensing the distance between them. They also perceive ambient light levels and have their own controllable colored light output. The research aim is to explore how really large numbers of fairly simple agents can work together to achieve a task.

A large number of Kilobots can be programmed simultaneously with open-source software and modern development tools.[15] For example,

 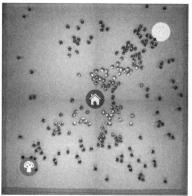

2.2 Left: A swarm of 900 Kilobot robots. Right: a vertical view of 200 Kilobots involved in a foraging experiment. Images courtesy of Sheffield Robotics, University of Sheffield, 2018 (left) and Andreagiovanni Reina, Sheffield Robotics, 2019 (right).

Kilobots can perform self-assembly into a shape. The desired shape, essentially a silhouette, is given to all the robots, and they track round the edge of their neighbors and self-locate to form the shape. An identical program in each Kilobot contains a few simple procedures, like edge following. Other examples of this global-from-local behavior include foraging for food sources and returning to a designated nest, and following a leader in a snake formation.

A nice example of a practical swarm application is the automated robot warehouse. There are many kinds of automated warehouse: Some use conventional robot arms for loading pallets, some have high-speed gantries that access enormous arrays of storage space, while others use small trucks that navigate via various high-tech means around the floor of a warehouse. The latter approach often employs swarms of mobile robots to fetch and collect products from dispersed storage locations to make up a specific order. Perhaps the prime example is the Amazon robot warehouse, where small modular robots are able to move *under* a crate, jack it up off the floor, and carry it to the target destination. It is entertaining to see the orchestrated motion of dozens of robots chasing up and down the main "roads," entering side alleys, and lifting up and putting down their heavy loads. They whizz about the warehouse but, of course, never collide with each other or with people.

These robots were produced by Kiva Systems and are the brainchild of Raffaello D'Andrea, one of the company's cofounders.[16] Kiva Systems was set up in 2003, and some 80,000 of its robots are in use; Amazon found these little robots so valuable that they bought the company in 2012. Initially, it seemed that Amazon was going to enter the robot supplier market, but that didn't happen; Kiva has been absorbed into Amazon and renamed Amazon Robotics. There are some websites that contain interesting details on how the robots are used.[17]

ENTERTAINMENT ROBOTS

Another, very different, field is *entertainment robotics*. This is a large and diverse market. One of the first entries into this realm was a robot dog, AIBO (whose name stands for Artificial Intelligence roBOt), which was marketed by Sony between 1999 and 2005. The great thing about toys and entertainment is that these products don't have to address a particular task; they just have to be fun to play with! AIBO was provided with software tools that allowed the owner to program their 'pet' with its own behavior and personality. AIBO was very sophisticated, with a powerful processor, memory, and a good range of sensors: camera, tactile sensors, heat sensor, range finder, acceleration and vibration sensors, microphone, and speaker.[18] It also had some built-in programs for useful functions, such as righting itself if it was pushed over. AIBO was a platform for education as well as entertainment; it led the way to introducing children to engineering and science through fun workshops with robots. The RoboCup robot soccer competition, previously mentioned, held a special Four-Legged League, in which teams of AIBOs tracked an orange ball with their head cameras.

Surprisingly, after a gap of 11 years, Sony has released a new version of AIBO in 2018. It seems that Sony carefully reviewed the previous product and enhanced all the positive features. The new AIBO is more rounded, less robotic, with big eyes and cute appearance and behavior. It has more tactile sensing (it likes "a pat on the head, a scratch on the chin, or a gentle stroke down the back"), better vision and hearing, and can recognize people and commands. It has a curiosity drive that allows it to explore its environment, and it remembers familiar locations. This is very much a social robot, but despite Sony's claims that it "develops relationships with people," there is still some way to go.[19] A helpful design constraint is evident in the fact that all AIBO's toys must be colored bright pink!

Another example of a modern entertainment robot is the table-top Cozmo from the AI aware toymaker Anki. The design of Cozmo is a lesson in how to make attractive, engaging toys for the commercial market. Cozmo moves on two tracks, facilitating nimble and reliable movement; has a simple lifting device, thus allowing objects to be pushed or carried; and has a small, front-facing screen that displays an expressive face. Together with a series of speech utterances, the face and animated movements allow a kind of communication. By restricting Cozmo to a table top ("Cozmo Prefers a Clean, Open, Indoor Space to Play,"[20]), the designers cleverly constrain the operating space and save some resources for games and further interactions with humans within this environment. Cozmo has a built-in camera with computer vision and machine-learning software, much of which runs on the user's smartphone. It can learn the layout of obstacles, be taught new actions, and gives "emotional" responses according to its internal state. It's not human, but it's small, cute, and feels alive. And even better, this little robot costs a few hundred dollars, many times less than AIBO. The success of Cozmo has led to a second-generation toy, Vector, with more abilities and better learning via a link to the cloud.

COMPANION ROBOTS

Another example of a social robot is Toyota's Kirobo Mini robot. The car company returned to robotics in 2013, when it sent its small Kirobo robot (34 centimeters tall) to the International Space Station to chat in Japanese and "become the first companion robot in space." They have now produced a version for sale in the general market, the Kirobo Mini robot, only 10 centimeters high, a companion robot designed to ride in cars and act as a conversation partner.[21] What the Kirobo Mini robot hears is sent to the user's smartphone, and then to a Toyota system running in the cloud. The cloud system recognizes words and orchestrates the conversation. Like the Keepon robot, the Kirobo Mini turns to look at the speaker and makes gestures to indicate simple emotions. The robot costs a few hundred dollars and is a signal of Toyota's large new research investment in robotics and AI, particularly for autonomous cars and social and care robotics.

A number of small companion robots are now entering the market. Some of these are essentially virtual assistants, like Siri or Alexa, based on a static or mobile plastic body with some Keepon-style animated type body motion. The business model for robot start-ups involves more risks than for software companies, and a number have failed; for example, the robots Jibo and Kuri are no longer available.[22] The initial high retail price is an issue for domestic markets. Other designs are based on simulated animal behavior, effectively artificial pets. Pets provide companionship for people with infirmities or limited mobility and have proven therapeutic value for people living alone. Robots such as MiRo are aimed at patients in hospitals and nursing homes, where live pets are not allowed. MiRo is mobile, the size of a small dog, with animated eyes and ears and a host of sensors (vision, hearing, tactile body sensors, accelerometers, and other features).[23] A well-known robot claimed to elicit emotional responses and reduce stress is Paro, an artificial baby seal. Paro is covered in fur and is thus more tactile than plastic, but it also carries a high purchase price.

HUMANLIKE ROBOTS?

There are now so many kinds of robots and so many application areas for robotics that it is impossible to cover them all; over 1,000 different types of robots are recorded in the A to Z listing of *Robotics Today*.[24] But the short overview given here shows that robotics can be applied to any application area, and the difficulty of engineering them depends upon how far the task can be simplified and how far their working environment can be constrained.

But now we ask our key question again: Do they think and behave like humans? They can certainly *behave* in ways to suit the task at hand, but they don't deal with their tasks in a human way. Commercial robots are highly engineered to perform very specific tasks well. Mower robots cannot be converted to cleaning floors, as similar as they are to floor cleaners, because they are designed specifically for grass cutting. Engineering is focused on achieving solutions by the most efficient and effective means, and usually this rules out human ways of doing things, as they are often not the most efficient.

Therefore, looking at most commercial robots, we see highly focused, task-specific devices that perform very well at the particular task they were designed for but are quite unable to show the flexibility necessary for addressing other tasks. The key human characteristic of adaptability, which gives us such versatility, is notably absent. It is our ability to cope with *all* kinds of situations under all kinds of risk, uncertainty, and ignorance that is central to human achievement. You don't need human versatility to do repetitious, simple, or boring tasks, and hence machines are much better for such jobs.

Only the entertainment/companion robots show any tendency to behave *like humans*. This is because the need for dialogue is built into their task specification—they are designed to engage and play with their human owners. All the other robots have primitive user interfaces because that's all they need to communicate basic commands and signals. Robots like Cozmo and Kirobo Mini have to respond and react to people and are provided with programs that sequence appropriate actions from common behavioral patterns. However, they still lack the flexibility of humans. They do not really understand interactive exchanges and don't learn or adapt through social contact. They offer simple simulations of behavior, but they do not think or behave like humans.

OBSERVATIONS

- Application areas for robotics are extremely diverse, covering all possible activities. This leads to a wide range of very different devices.
- The physical form of a robot determines what it can do. Conversely, a given task may require a particular anatomy. Here's a crude example: Wheels are excellent for flat surfaces, while legs are better for rough ground and stairs.
- Robots are designed for particular tasks. An engineering task specification usually defines the function and structure of a device or machine.
- Constraints on the working environment make the problem simpler, and thus easier and less risky, and they allow for less costly solutions.

- The downside of good engineering is that the more efficient robots become for performing one particular task, the less flexible they are for doing anything else. Consequently, most robots do not perform at all outside their task specification. There are no general-purpose robots.
- Software is often embedded in robots, but remote links increasingly are being used to access very large data sources or complex AI systems.
- Social robots, mainly intended to engage with humans, are much less task-centered and will need to be more open-ended.

3

FROM RESEARCH BENCH TO MARKET

Which single word best characterizes science?
—Curiosity
Which single word best characterizes engineering?
—Creativity

Commercial robot offerings are not as remarkable as the new software coming out of artificial intelligence (AI) companies. Of course, robotics and AI are not the same thing, but robots increasingly adopt and use the new AI advances. So let's continue with our constraint assessment idea and look at some research robots.

Research robots are usually more sophisticated than commercial robots and are capable of demonstrating some very impressive behavior, but being experimental systems, they are often incomplete or dysfunctional in some nonvital way. It is useful to consider the research process model, which is rather different from that of much commercial development.

Where do robotic ideas originate, and how do they get tested? What are the drivers of innovation? Most commercial robotic products have their origins in upstream prototypes in robotics research laboratories. Such prototypes are usually not directly motivated by the marketplace, but rather are driven by curiosity and the search for new knowledge and understanding. Scientists are at their most productive when they explore new ideas

3.1 The Honda P3 humanoid robot, an early precursor of the ASIMO robot. This walking and stair-climbing robot stands 1.6 m (5.25 ft) tall and weighs 130 kg (286 pounds). Image courtesy of Honda Motor Corporation.

and try to gain a better understanding of basic scientific principles. Such principles are the bedrock of most of the engineering and technology that we all rely on. The downside is that they may sometimes find themselves in blind alleys, or their results may turn out to be negative or inconclusive, so their work does not always lead to new advances that can be immediately exploited. But this is a necessary condition of science. In fact, it is essential for the creation of an environment that can support breakthroughs and radical new advances.

But this creativity comes at a cost. For example, the Honda development program for its P3 humanoid robot (figure 3.1) took 10 years, and the estimated cost was over $10 million. The researchers started on legged locomotion with bipedal designs in 1986 and worked through a succession of improved experimental designs,[1] producing the P3 in 1997. This research-and-development (R&D) program paid off, as it provided the groundwork for the development of the impressive and world-famous ASIMO robot.[2] (But even for ASIMO, the payoff was not in profits from sales, but in advertising and public relations.)

So research is expensive, and it can be risky, but it is essential to make progress. The high costs of research have to be recouped in future market returns; often only large companies can afford this level of investment. Even the largest companies sometimes decide the R&D time is too long for continued investment; for example, both Sony and Microsoft canceled their robotics programs after several years. Microsoft had a short adventure, starting in 2007 when Bill Gates launched an ambitious program promising "a robot in every home," but the robotics group closed down only seven years later.[3] Sony's foray into robotics was quite successful in technical terms. Its 61-cm (2 ft)-high QRIO humanoid robot was an agile mover that could walk, run, and dance and had many attractive features, but it was never sold, and Sony's robotics activities shut down in 2006.[4]

Hence, many companies don't want to start from scratch, and at any rate, it is not sensible to try to drive innovation by agenda; most advances are serendipitous and their emergence is not controllable. It is common nowadays for smaller, innovative companies to be bought by the large corporations after they've made a significant research breakthrough that needs investment to bring to market. Consequently, companies harvest ideas, technologies, and know-how from universities and other government-funded institutions, where basic research is carried out more efficiently and at better value-for-money ratios. University spin-off outfits and other small, specialist start-ups are often bought by bigger companies, as seen in the race to acquire high-technology inventions, especially robotics and software, over the last decade.

This picture of expensive equipment, laboratories, and funding budgets suggests that robotics research cannot be done by individuals, unlike the legendary computer company start-ups that began in a bedroom or garage.[5] As a result, most *basic* science is not funded through industrial companies (but there are some notable exceptions, such as Honda), but instead scientists sustain their work by bidding for funding to governmental and international agencies. Unfortunately, the politics of the day often influences the agenda of these funding agencies, so we sometimes see "fashions" occurring in science. These are marked either by highly publicized breakthroughs, attracting disproportionate levels of funding, or a so-called funding drought, where negative results or controversies can cut off all funding. Ironically, some topics are rediscovered and come back into favor several decades later: Well-known AI examples include computational language translation and artificial neural networks (ANNs). Meanwhile the big four (Google, Amazon, Apple, and Microsoft) set the agenda for customers, as they design and select new and future products for the market.

So commercial robots are built on the back of robotics research and depend on good ideas being proved and offering good potential as products. The selected good ideas then undergo extensive programs of development, which are funded via investments by companies that see profit-making potential.[6] By looking at the current state of research, we can get a hint of what advances will enter the development pipeline and what kind of products we might expect to see in the near future.

BIN PICKING

Robotics research nowadays covers every area imaginable, so I have selected a few current topics. Up to now, we have seen how Amazon has found an effective way of using swarms of mobile robots to automate the movement of goods and customer orders around its warehouses. But it still needs people to pick the products out of the bins and assemble individual orders, and this is clearly Amazon's next target for automation. Since 2015, the company has been organizing an annual Robotics Challenge. In these events, teams from around the world compete to build robot demonstrations of pick-and-place tasks that essentially try to reproduce human skills in retrieving specific objects from a jumble of different items in a box. This is known as working in an "unstructured environment," and the difficulty is demonstrated in figure 3.2, which shows a typical batch of mixed household products. Notice that items have to be removed (picked) in a given order, which is much harder than just removing the easy ones first.

In 2017, sixteen international teams competed in the Amazon Robotics Challenge; eight reached the final, and the first-place winners were an Australian team from Brisbane.[7] Amazon described the challenge as follows: "The 2017 Amazon Robotics Challenge is a skill challenge sponsored by Amazon Robotics LLC to strengthen the ties between the

3.2 A bin-picking task. Selected items are to be removed individually from a box of jumbled and very different products.

industrial and academic robotic communities and promote shared and open solutions to some of the big problems in unstructured automation. The Challenge will task entrants with building their own robot hardware and software that can attempt simplified versions of the general task of picking and stowing items on shelves. The Challenge combines object recognition, pose recognition, grasp planning, compliant manipulation, motion planning, task planning, task execution, and error detection and recovery. The Robots will be scored by how many items are picked and stowed in a fixed amount of time. Teams will share and disseminate their approach to improve future Challenge results and industrial implementations."[8]

Once a prototype solution has proved feasible and reliable, Amazon will implement it across the organization, thus gaining completely autonomous warehousing and massively reduced costs. This will repay many times over the few hundred thousand dollars the company has spent on the Challenge. This illustrates why technical challenges and competitions are popular with sponsors; they are a very effective and cheap way of stimulating innovation.

BIOROBOTICS

Much general inspiration for robotics research comes from the animal world. *Biorobotics* involves the selection of particular animal characteristics and behavior to reproduce through experimental implementations of the underlying mechanism. Examples that have been investigated include various birds and mammals, snakes, bats, and many forms of insects and reptiles, where locomotion, body dynamics, and specialized sensing regimes have all been studied. Robot fish include a robot tuna from the Massachusetts Institute of Technology (MIT) (Triantafyllou and Triantafyllou, 1995), an extinct coelacanth from Mitsubishi (Terada and Yamamoto, 2004), and three robotic fish from Essex University that swam in the London Aquarium (Hu, 2006). An interesting line of research at Vassar College is using robotics to understand the mechanics and dynamics of the spinal column of sharks, blue marlin, and tadpoles. The elasticity, stiffness, and other properties of the vertebral column in sharks are highly tuned by evolution for rapid and acrobatic swimming. But before the vertebrae had evolved, the earliest ancestors had backbones made from a solid flexible rod known as a *notochord*. It is not clear how these evolved and how the various biomechanics affected swimming performance, so the team at Vassar has been building robotic versions and measuring their performance while varying the mechanical parameters. Various arrangements of backbone give different energy storage properties and hence change dynamic behavior. This is another use of robotics: to model biological systems to understand the biomechanics of a current species and figure out how extinct species behaved.[9]

CARE AND ASSISTIVE ROBOTS

One of the big political worries for nearly all governments of the developed world is their aging population. Falling birth rates and increasing life spans mean that in the near future, there simply will not be enough younger people available to care for an enormously expanded population of older people, many with age-related impairments or disabilities. This is not an assumption, but a fact that we must face, demonstrated by demographic analyses of current populations.[10] Also, extended families are contracting in size and work patterns are changing for young people, exacerbating the growing shortage of caregivers available to provide support for aging populations. For these and other reasons, the care equation is going to become a crisis without careful planning. Over the last few decades, this concern has been a driver for supporting and funding robotics research in the social and care sectors. Governments and economists have come to believe that technology might provide a solution and have been motivated to fund large national research initiatives (e.g., in the US, Europe, and Japan).

There are two main care scenarios: helping people live independently in their own homes for as long as possible, and care homes, where people live together and centralized services can be provided. The former is the most desirable, from both the well-being and economics standpoints, but it is also the most difficult, as it involves dealing with every possible type of home and highly personalized needs and requirements. Indeed, this is perhaps the most taxing application area for robotics because so many different functions may be needed, and enormous flexibility will be required to adapt to various home scenarios. We have mentioned service robots before in this book, but they mainly transport items and deliver them inside hospitals or companies, whereas *assistive robotics* must support people by performing a wide range of tasks. Human caregivers typically visit their clients several times a day and help with getting dressed, preparing a meal, tidying the home, dealing with any problems, and then assisting with going to bed. It is difficult to imagine any current robot system being able to do some of these tasks, never mind all of them.

AFFECTIVE COMPUTING

Trying to build a completely humanoid robot that could reproduce the behavior of a human caregiver is *not* a sensible approach. So researchers are investigating various home tasks separately, including identifying and welcoming visitors; navigating around the rooms of a home; and finding and retrieving personal objects. Interactions with the user will be absolutely crucial: A good rapport and fluid communications are essential for acceptance, and this is obviously an area where social robots are needed.

Much work is underway in the specialized field of research investigating *human-robot interactions*. This covers the understanding of natural speech, dialogue and conversation, gesture and other forms of nonverbal communication, emotional behavior, and expression of meaning and intention. This is a very active research area and is producing valuable results thus far. For example, autistic children have shown responsive preferences for robot toys over human caregivers, and this stimulation and support from robots has enhanced their development (Wainer, Dautenhahn, Robins, and Amirabdollahian, 2010).

A particular topic is the study of empathy and human affective states. This is known as *affective computing*, or *emotion AI*, and concerns both the generation of emotional responses and the recognition of emotions in humans. Visual facial expression software can fairly accurately estimate human emotional moods by detecting changes in the facial muscles (McColl, Hong, Hatakeyama, Nejat, and Benhabib, 2016). Psychologists classify a range of emotions, but seven major categories are easily recognized by these systems: sadness, happiness, anger, fear, surprise, contempt, and disgust.[11] Emotions can also be detected from speech passages, or even textual material. Research robots have been built that can produce these expressions of common human emotions as a communication aid. Clearly, empathy is very important for social robots, but we will consider later (chapter 14) to what extent emotions are necessary.

HUMANOID ROBOTS

The most ambitious level of biorobotics is humanoid robots, where the aim is to implement models of the human form. Of course, these are the most overhyped kinds of robotics research; the image of a humanlike device raises expectations that it will exhibit humanlike behavior in every other way. There currently exists an enormous range of humanoid robots, ranging from build-it-yourself hobby kits, to affordable, multifunction commercial products, to expensive laboratory models used primarily for research. Several reviews are available,[12] and we will look at a few research milestones here.

3.3 The Honda ASIMO robot. Image courtesy of Honda Motor Corporation.

As mentioned previously, Honda has a long history with humanoid robotics. When the research started, just standing on two legs and walking without falling over was quite a challenge for humanoid robot research! Following the E series of prototypes from 1986 to 1993, the P series was developed in 1997.[13] This was a laboratory robot (see figure 3.1), not a product, but Honda cracked the complex control problems of walking and foot dynamics, which marked a significant breakthrough.

Based on this work, the smaller, 1.2-m (4-ft) ASIMO robot was produced from 2000 onward (see figure 3.3). This had even better flexibility and dynamic agility and has become known worldwide as a milestone in body mechanics for humanoids.[14] A new model was launched in 2011; ASIMO can now run forward (up to 9 km/h [5 mph]) and backward, jump, and hop on one leg, walk over uneven surfaces, climb up and down stairs, recognize faces and voices, and aim and kick a soccer ball. These advances are due to the robot's much improved postural and balance control, allowing fast response to disturbances and local terrain variation. ASIMO is also able to detect people and predict their path of movement so as to modify its behavior to fit in to and better interact with the local human environment. Honda's latest robot, the E2-DR Disaster Response Robot, is a search-and-rescue humanoid with a thin profile and extended joint flexibility so that it can squeeze sideways through gaps or holes.[15]

Other companies include Toyota, which, in 2005, developed a walking humanoid that played the trumpet (using artificial lips and a supply of compressed air), swayed its hips, and tapped its feet in time to the music. This was typical of most humanoid research; it was an exploratory research project, and only one or two robots were built. But a few companies aimed at commercial markets from the outset. SoftBank Robotics (formerly Aldebaren) produced the small 58cm (2ft), relatively affordable NAO robot ($8,000) in 2006, and over 10,000 have been sold worldwide, mainly for research and education. The company has grown to more than 500 employees, and its Pepper personal robot, launched in 2014, sold more than 10,000 units in three years.

The dynamics and stability problems associated with walking robots have been largely solved as a result of these research programs. For example, the Atlas humanoid robot, produced by Boston Dynamics, can withstand people pushing it with a pole, knocking it sideways, and even causing it to fall flat. Boston Dynamics (a spin-off of MIT) is well known for its range of four legged, cargo-carrying robots ("dogs" and "mules") that can scramble over the roughest terrain and even stagger and recover when on ice or when kicked.[16] Google bought Boston Dynamics in 2013, but after three years, it put the company up for sale (it is now part of the SoftBank Group).

In the realm of *space robotics*, the National Aeronautics and Space Administration (NASA) has been investigating very sophisticated robots for performing tasks in space. NASA has a history of adventurous trials of AI and other advanced technologies, probably because most space work is for one time only and experimental in its very essence. For example, in 1999, the AI software system, Remote Agent, was given sole control of the space probe, Deep Space 1, for several days.[17] Also, the agency's Robonaut is an upper-body humanoid designed for working with tools on the externals of spacecraft,[18] and its latest robot, Valkyrie, is an early prototype of a robot suitable for going to Mars.[19]

The roles of these robots include manual tasks, such as using hand tools to remove and replace components on the external surfaces of space vehicles and space stations. The humanoid anatomy allows the performance of actions analogous to human actions, facilitating flexibility in performing tasks in different ways, and human crews can more readily oversee the work and empathize with and appreciate any performance variations involved.

Closely related applications, but more down to earth, are *human helpers*, such as the robot in the SecondHands project. This is a European collaboration with the aim of providing support to maintenance workers during physical tasks in a range of situations.[20] When working on a car engine, or fixing something when up a ladder, one frequently finds oneself asking anyone nearby, "Please pass me that wrench/screwdriver/replacement light bulb." The SecondHands robot acts as a builder's assistant; it can pass things, hold things, and generally be helpful (see figure 3.4). Unlike human asistants, it can hold things for hours–literally–so "Keep your finger on that" is a command that a robot can carry out while the builder goes on a break—even if it lasts overnight. This example illustrates the value of a humanoid anatomy for collaborating robots—we can empathize with such familiar bodily forms and so more easily anticipate their behavioral abilities and limitations.

It is noticeable that robots are "coming out of the cage," both literally and metaphorically. The first generation of industrial robots was always contained within cages to protect factory workers from accidental injury, but research has allowed this constraint to be removed, and modern devices today are able to work in close cooperation with humans. For example, the Baxter robot from Rethink Robotics[21] was the first two-armed industrial robot designed to work in conjunction with people. It shows its internal state of operation in terms of facial expressions on a display screen, and it can be trained by guiding the arms through the desired positions and trajectories, which then can be played back. In the first two years, over 3,000 Baxters were sold.

Other humanoid designs have been developed especially for use in research projects. In chapter 13, we will meet the iCub robot, which is helping research laboratories explore models of infant behavior.

3.4 The SecondHands robot. Photo courtesy of Ocado (c) 2017.

WHY HAS INDUSTRIAL ROBOTICS BEEN SO SUCCESSFUL?

Industrial robotics is nowadays less of a research area and more of a fully developed commercial proposition. Many technical problems have been solved, and all kinds of robot arms and mobile devices are used in manufacturing industries worldwide. This is now a mature market, with an annual value of $32.8 billion in 2018; more than doubled in five years. Of course, research continues on manufacturing technology, such as sophisticated inspection techniques using vision, and adaptive programming methods to deal with flexible production runs. But the strides made in the factory have stimulated robotics research to move on to the more challenging problems found in less constrained environments.

The secret of the success of the modern, highly efficient manufacturing plant rests on the well-organized structures that underpin everything. The combination of physical organization and computer-based sensing of all objects and processes allows nearly everything to be controlled and monitored from factorywide software. Information is captured and held by computers that know all about the materials, tools, and components as the product is being built. Everything is tracked and accounted for. So the spatial location of each cake in a cake factory, as well as a range of parameters of its current state, is measured and analyzed, and any errors are dealt with quickly.

The key for integrating robots into manufacturing systems is this structured environment approach. The factory is an engineered environment, designed and controlled. The robots can access a wealth of online infrastructure data and determine what they are required to do: move a component from one machine to another; fit a car windshield in place; or paint chocolate icing onto a batch of cakes.

However, as soon as we get out of the factory, we hit all the variability, uncertainty, and sheer volume of stimulating possibilities that the real world offers. Such unstructured environments make matters much more difficult for a robot, not least because there are no readily available data on these environments and no indication of how best to deal with them.

In other words, as the degree of structure in the robot's environment decreases, the degree of chaos increases and life gets much more difficult. We can characterize the level of chaos in terms of four features. *Uncertainty* refers to the degree of doubt or unreliability likely to be found in the environment; this will indicate how much sensing will be required. *Novel events* are unexpected occurrences and situations that have not been experienced before, and therefore cause problems and require extra work in order to recover from error situations. *Available knowledge* concerns the amount of prior information that can be used to design a system for a task and/or be directly utilized by the robot. This might be knowledge of natural physical properties, such as gravity or sunlight, or it might be human-made rules, such as road markings and driving conventions. *Order* refers to the degree of regularity and structure found in the environment, ranging from precise tidiness to random piles of junk.

Using these features, we can estimate their value for some of our robot examples (see table 3.1). I use seven qualitative values: negligible, very low, low, medium, high, very high, and maximum.[22]

We can summarize the total chaos in these environments by mapping the qualitative values onto a numeric scale and adding them up.[23] The result is an overall chaos score for each application, given in the penultimate column in table 3.1.

The table shows that the factory situation is the most accommodating environment for robot applications because design engineers have so much useful information and the environment itself can be engineered to be benign. Lawn-mower robots also have constrained working environments (they simply don't work anywhere else), but they still have to sense any obstacles or disturbances. Warehouse robots transporting pallets have well-defined routes and tasks, but humans can get in the way, so this is just slightly more chaotic.

Because the practice of medicine is so safety-conscious, robot operations are always supervised by humans, and the chaos factor mainly arises from biological variations. The milking robot has slightly more uncertainty in dealing with variations in behavior from different cows, although the rest of the process is well controlled. Service robots have to navigate around buildings and internal obstacles, as do security robots. But security robots are required to deal with more unexpected events.

Agriculture covers a wide range of problems, and some tasks are already satisfactorily automated (e.g., combine harvesters). Here, we are considering crop monitoring, weeding, and picking; these functions are not yet out of the research labs—but almost.

Simple companion robots, here classed under "Entertainment," potentially have very complex environments, but the complexity is usually reduced by restricting the scope of any dialogue or interactive behaviors.

Autonomous vehicles face differing levels of difficulty, depending on where they are driving. Open highways, with wide lanes, clear markings, and no pedestrians, are less complex than off-road locations, where rocks and trees must be avoided. City-center driving is the most hazardous, with cyclists, pedestrians, children, subtle road signs (like "one-way street"), emergency vehicles, and other complications. This is where more difficult driver action is required, such as reversing in order to free up

Table 3.1 Assessing the difficulty of robotic applications.

Application	Level of chaos				Chaos score	Task difficulty
	Uncertainty	Novel events	Available knowledge	Order in environment		
Factory robot	Very low	Very low	Very high	Very high	8	1
Lawn mower	Low	Low	Very high	Very high	10	1
Warehouse—transport	Low	Low	Very high	High	11	1
Medical robot	Low	Low	High	High	12	2
Milking robot	Medium	Low	High	High	13	2
Service robot	Medium	Low	High	Medium	14	3
Security robot	Medium	Medium	High	Medium	15	3
Self-driving—open highway	Medium	Medium	Medium	Medium	16	4
Agricultural	Medium	Medium	Medium	Medium	16	4
Entertainment	Medium	Medium	Medium	Medium	16	4
Self-driving—off-road	High	High	Medium	Medium	18	5
Warehouse—bin picking	Very high	Low	Low	Low	19	5
Self-driving—city center	Very high	Very high	Medium	Low	21	6
Home care	Very high	Very high	Low	Very low	23	6
General-use humanoid	Very high	Maximum	Very low	Negligible	26	7

a jam at an intersection. It might be surprising that bin-picking and home-care robots have such high scores, but like driverless cars, these are problem areas with a great deal of real-world uncertainty.

Finally, humanoid robots get the highest chaos score because we are assuming that these are general-purpose machines, and thus they will have to deal with virtually unknown environments, as well as people, with all the novelty, interactions, unknowns, and disorder that entails.

Most of the examples in table 3.1 are well defined, but there are four application areas that are more difficult to pin down: medicine, agriculture, entertainment, and home care. These are really categories of task, not task definitions, and they contain many subareas, with difficulties ranging from fairly easy to very hard. For example, the task specification for harvesting fruit will be much more difficult to implement if the fruits have to be picked individually, rather than shaken off a tree! As we are looking at research challenges, the table assumes the highest levels of difficulty for each of these application categories.

Of course, tasks can sometimes be easy, even in difficult environments. For example, a monitoring robot might maintain the sensors for measuring some local variables: temperature, visibility, pressure, and pollution without being required to perform any very complex actions. So we can look at overall task difficulty as another (rather crude) way of measuring and classifying robot applications. Using a 7-point scale, the last column in table 3.1 gives a rough assessment of task difficulty for these examples. Not surprisingly, difficulty increases with chaos, as tasks often become more complex and ill defined. The humanoid robot is still given the highest rating, as potentially, it has to perform a huge range of tasks: that's what *general-purpose* means.

THE CURRENT STATE OF ROBOTICS

From these examples, we might get the impression that robots can do practically anything. They can vacuum the carpet, work in outer space, swim, fly, act as personal assistants, imitate biological behaviors, automate industrial factories, and be fun to play with. In fact, any task can be seen as a potential robot application, and probably has already been tried in some laboratory somewhere (and, yes, robots can now play table tennis in real time[24]). There just isn't space to list the huge diversity of robot designs, abilities, and applications. It seems that most robotic problems either have been or will be solved in the near future.

The most obvious feature of the collection of robots briefly reviewed in this chapter and the previous one is that they seem to have no physical features in common. They differ markedly in their anatomies: They might have legs or wheels; they may or may not have arms or hands; they may be copies of particular animals. The physical form of a robot is open to all possibilities, but whatever their design it will be closely linked to the task or purpose of the device. Just as evolution emphasizes the link between an animal and its environmental niche, so designers choose appropriate bodies and mechanics to suit the application. For example, snake robots have been developed for inspection tasks deep inside pipes in nuclear reactors.

Regarding innovation in robotics, the proven way of stimulating progress is to organize a challenge or competition, such as the Amazon example described previously. Challenges can be found for all aspects of robotics (from home-care devices to robot sailboats); they often run in annual series, some offer large cash prizes, and they can be irresistible to research scientists, as well as stimulating amateurs, schoolchildren, and other budding engineers. They range from serious, very difficult challenges that only allow well-funded teams to enter, to open, inclusive events that involve cheap, even handmade robots.

We have seen how the RoboCup competition stimulates progress through the theme of robot soccer. This has been running for 20 years and is established as a worldwide event with many levels (leagues) of sophistication. A good example of a high-end challenge is the US Defense Advanced Research Projects Agency (DARPA) challenge series. From driverless cars (2004 and 2007) to recovery and rescue robots (2012 to 2015), DARPA has put big money into stimulating innovation.[25] Most challenges start off with failure but soon make progress as they stimulate human ingenuity. And, as Amazon demonstrates, challenges are a very efficient way for companies to invite solutions for their own problems.

Established robots, in factories and companies, have not yet been very human friendly, and they didn't need to be. Why should they, if they can just do their job? But it is clear that human-robot interaction is going to be much more important in future devices. The area of social applications is undergoing intensive research, but we can already see a few characteristics that will determine a robot's acceptability. Small, cute robots are attractive, while large, powerful machines are perceived as threatening. The large and heavy Honda P3 raises anxiety levels, while the smaller ASIMO is seen as charming and amusing. The other examples we have seen thus far in this book, Keepon, Cozmo, and the Kirobo Mini, are all small, very cute, responsive, and make humanlike movements and gestures. They also talk and recognize speech and human movements, as do many new robotic products. It is clear that for social robotics and areas like home care, where long periods of companionship are expected, these kinds of interactions will be essential and need to be developed to a high level.

The field of social robotics holds regular conferences and publishes journals like the *International Journal of Social Robotics*, which covers research into conversational systems, emotional expression and other nonverbal communication, and interactive analysis. While this technology makes important contributions to our quest for human friendliness, our main concern is the broad view of learning and thinking algorithms that robots require to behave like humans. Of course, conversational systems are now widely used on phones and other domestic devices, and these could easily be added to robots to make them more user-friendly. But there is much more to social interaction than semiformal dialogue; we will give more attention to these systems in chapter 8.

Apart from a few lab-based experiments that can demonstrate specific humanlike features or abilities, there are still no robots in existence that get anywhere near general human behavior. Despite many decades of scientific work and engineering experiments, we still do not have a fully satisfactory answer to this problem. General-purpose behavior is elusive. For example, extremely lifelike robots have been made of the human form, but as soon as they move or talk, it becomes obvious that they are not authentic.[26] The key quality of being *humanlike* requires more than looking human. Indeed, robots that look too much like humans create unease.[27]

On the positive side, we can expect better robot behavior in the future, and many more individual and exciting designs, because physics, materials science, and mechanical engineering are all producing more new materials and methods that deliver new functional properties and possibilities. In addition to better manipulation and dexterity from more advanced robot hands with tactile control, we can expect soft-bodied robotics to interact with objects in safer and more flexible ways. There remains much room for exciting advances in robotic hardware.

Having considered some hardware, it is now time to take a brief look at software. Software is important because it can create new functions for the existing robot hardware. Thinking allows humans to find ways of solving problems better than robots, despite their less flexible bodies and limited physical powers. Robots must also "think" at some level, and in the next chapter, we turn to AI to explore this vital issue.

OBSERVATIONS

- Robotics research is much more expensive than software research, requiring fully equipped hardware laboratories and longer development cycles.
- New innovations for commercial robots depend upon funded research in robotics and software (i.e., artificial intelligence).
- Innovation tends to emerge from small teams. Challenges and competitions are very cost-effective ways of stimulating progress.
- The degree of chaos in a robot's environment and the complexity of the intended task are major factors that determine the difficulty of building a successful robot. Any restrictions on the working environment or simplifications of the task will greatly help to reduce the designer's difficulties.
- Human-robot interaction is a key topic because robots will increasingly work in cooperation with humans. Acceptable levels of social interaction will require robots to behave like and at least appear to think like humans. Small and cute robots are far more attractive for human interactions than large machines.
- Social robots will need to interpret human emotional states, display attention and engagement, match their gestures and idioms to human norms, and identify and remember their human contacts in considerable detail.

- Research on social robots is ongoing, and many components are being researched individually, but the work is fragmented and social interaction is still primitive. There are still no robots in existence that get anywhere near general human behavior.
- We need to distinguish *intelligent robots* from *humanlike robots* that are intelligent in a particular way: They behave like humans. AI will drive many amazing robots, but they will be designed for specific classes of problems or tasks.
- Humans are actually general-purpose agents; they learn to deal with many situations and can adapt their skills to cope with all kinds of problems. AI has always struggled with the general-purpose idea.

4

A TALE OF BRUTE FORCE

An algorithm that once took 10 minutes to perform can now be done 15 times a second. Students sometimes ask my advice on how to get rich. The best advice I can give them is to dig up some old algorithm that once took forever, program it for a modern system, form a start-up to market it, and then get rich.
—Maurice V. Wilkes, in Denning, 1999

Artificial intelligence (AI) is a branch of computer science that is concerned with finding ways that computers can perform tasks that are considered to require intelligence.[1] Most people have no trouble with the concept of intelligence and take a "I know it when I see it" approach, and there is something to be said for this. After all, we don't need formal definitions to assess the skill of a person playing a sport, or the performance of a musician—we can all appreciate the skill of a performer. However, *intelligence* is actually a rather vague and ill-defined term, and theoreticians and researchers feel the need to describe their work more precisely.

We might try to define intelligence to say that an AI system shows intelligence when it demonstrates some form of cognition, or thinking. But *thinking* is another ill-defined concept, and we would then have to define this more exactly. The problem here is that all the ideas that spring to mind to characterize intelligence, such as understanding, reasoning, learning, and acquiring and applying knowledge, are themselves similarly difficult to define; we end up producing very long, complex chains of ever more diverging subdefinitions. In fact, two researchers have found and listed 71 definitions of intelligence (Legg and Hutter, 2007). Trying to define these rather nebulous concepts formally means entering the realm of philosophy, but we will avoid going down that route here.

It is interesting that in 1950, the great computer theorist and pioneer Alan Turing also considered this problem. Turing is well known for his cryptography achievements in World War II, but he was a brilliant mathematician who applied his talents widely, covering logic, philosophy, biology, and the formal theory of computation. Generally recognized as the father of computer science, he was also a pioneer in the area of AI. He addressed the problem of trying to define AI in words by replacing definitions with a kind of test by which computers have to pretend to be a human well enough to fool an interrogator into believing that she or he is actually dealing with a human. Turing predicted the test would be passed by the end of the twentieth century, but this did not happen, and it is proving to be a very difficult challenge (but more on this later in chapter 8). However, Turing also believed that the term *thinking* would eventually become widely used and accepted: "[by the year 2000] …one will be able to speak of machines thinking without expecting to be contradicted" (Turing and Copeland, 2004, 442).

SEARCHING THROUGH THE OPTIONS

There is much confusion, controversy, and obfuscation surrounding AI. AI takes many forms and involves many very different techniques, but there is a common underlying motivation and challenge behind all AI systems. I want to focus on these general issues and clarify the important points. We need to look briefly at an AI example in order to build up an appreciation of how it works. So this chapter explains how searching techniques (a cornerstone method in computing) are used in the context of game-playing (the most studied application area in AI).

Consider the problem of working through a series of options. For example, we often have to make plans; to do this, we choose a sequence of available actions to get from a starting state to reach a desired state.

A TALE OF BRUTE FORCE   57

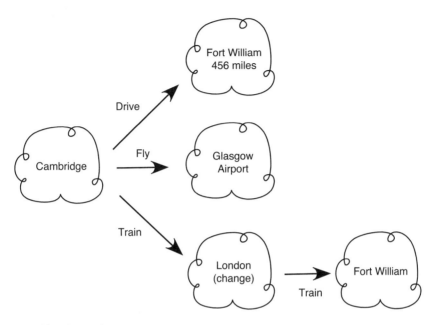

4.1 Planning a trip.

This might be to plan a route for a complicated journey, plan an event like a wedding, or decide how to organize a major refurbishment. Planning software is widely used in manufacturing, for organizing production processes; in commerce, for scheduling transportation; in education, for organizing room allocations; in business, for supply-chain planning; and in space, for organizing the extreme complexity of space missions. These are all *searching* problems because they involve sifting through many options to find the best sequence of actions or events.

The fundamental process is very simple. We begin in a starting situation, and we know of a range of actions that can be taken, each of which will move us to a new situation. If we don't know which action is best (as is often the case), then we can try any one of the actions randomly. If we don't like the look of the new situation, then we can ignore that one and try another. But if the new situation looks promising, then we select a subsequent action and create another new situation. We can keep doing this until we find a situation that satisfies our needs.

Figure 4.1 illustrates this process for Gwen, who lives in Cambridge, in the UK, and wishes to take a vacation in the Scottish mountains. She chooses Fort William as her destination, a good center for exploring the UK's tallest mountain, Ben Nevis. She has three possible actions: to drive, fly, or take the train. Considering each action in turn leads to three alternative possibilities (see figure 4.1). Each situation will have associated information, such as the time it takes, cost, and personal values such as convenience or preference. The driving option involves traveling over 450 miles, and Gwen rejects this as too tiring, so she does not consider it further. The result is similar for flying; the nearest airport is still a long way from the destination, so she rejects this option, too. The train option looks promising, but Gwen realizes that she has to go (backwards) to London to catch the best train (which does have the attraction of a luxurious overnight sleeper service). However, this is a through service with no changes of train required. This takes two steps to reach the desired result.

There are just two parts to this planning or searching process. The generation of the situations is a mechanical procedure and is often easy to program. We can call this the *generator*. The other necessary action is to *judge* each new situation: We need a way of computing when we don't like the look of a situation and when a situation satisfies our need. This is needed to both guide the planning and to decide when to stop. We can call this the *evaluator*.

This latter part is not so easy for a computer because it is application specific; it has to compute the personal values that we place on various situations. The whole process runs in a big loop: Generate some new situations and then evaluate them, generate more from those with the better values, evaluate them, and keep going until one with the desired criteria is reached.

THE WORLD CHESS CHAMPION IS A COMPUTER—SO WHAT?

Real life is complicated, so AI has always looked for simplifications (constraints) to make things easier for testing new ideas and methods. Game playing has been used for planning because games are much simpler than real life (they are defined completely by a few rules), so they are much easier to automate and understand. Consider computer chess, a classic and long-standing topic in AI. You may be wondering what chess has to do with robotics, but bear with me: Game playing is a theme that leads directly to a major breakthrough in 2012, and then onward to the current storm of AI activity.

Ever since the first electronic computers could be programmed, people have considered implementing computer chess as an interesting challenge. Alan Turing made major original contributions to this too.[2] Chess requires quite a bit of thinking to play well, so it appears to be a good test of intelligence. A simple program to decide which move to make next is actually not too difficult to design; indeed, it makes a good undergraduate programming project.

Of course, such exercises do not produce very high levels of play, but the basic search algorithm, as described here, is fundamentally the same as that employed in world champion computer chess programs. As for planning, the basic algorithm for chess-playing programs consists of a *generator* and an *evaluator*, as defined previously. These are outlined in boxes 4.1 and 4.2.

The farther down the tree of moves you manage to go, the more implications of the moves at the top will be seen, and the best one will be your

Box 4.1
How chess is automated—The tree of moves

> The Current Board position is the starting situation, and the actions are the various alternative legal moves you can make from there. Figure 4.2 shows a diagram of this process, known as a *search tree*. For illustration purposes, I've restricted the number of available moves to three. The three alternative board configurations resulting from the start are placed on the first level down. This is the computer's move. Next follows the opponent's reply, giving nine possible situations for level two, and then the computer plays again, and so on.[3] This is the *generator* at work, producing the search tree.
>
> The number of board situations grows very rapidly as the levels increase. For example, if we continue this tree with just three moves at each position, then at level 10, there would be over 59,000 possible board arrangements. With just five moves at each position, level 10 would have over 9 million possibilities! An average for chess is around 10 to 20 possible moves, and so the growth rate of the tree is truly extraordinary. This type of rapid expansion is known as *exponential growth*, or, in terms of the number of board combinations being generated, a *combinatorial explosion*.[4]

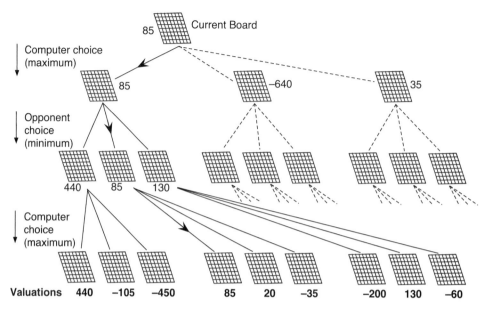

4.2 A two-person game search tree.

chosen move. It is known that master chess players do a lot of searching, so if a computer could search a bit deeper than the master players, then it should be able to beat them—or at least that's the theory. Writing chess programs is fun and very engaging, requiring multiple programming tricks to keep the process efficient and not waste valuable thinking time. Many creative people have worked on computer chess (and many other games), usually in their spare time. One attractive element of this enterprise is that measuring performance is easy, as all that really matters is the outcome of the game; therefore, the same ranking system used for human chess players can be applied to computer programs. In 1970, the first computer chess tournament was held, and from then on, regular tournaments have been a way for programs to be tested against other programs, and hence to measure progress.[5]

After many years of development, a computer eventually beat the world champion. In 1997, Garry Kasparov was defeated by Deep Blue, a machine that contained IBM processors using specialized chess hardware that could examine 200 million moves per second. Deep Blue began as a

Box 4.2
Evaluating the possible moves

> When as much of the search tree has been generated as possible (as limited by available time and memory), all the positions at the deepest level in each branch have to be assessed. An *evaluator* gives them values according to how good they look—positive numbers for the computer, negative for the opponent, and zero for neutral situations. These are marked "Valuations" in figure 4.2. Next, for each branch, the computer selects the best value if it were the computer's move, or the worst value if it were the opponent's move, and passes this value back up one level to the board above (see figure 4.2). This process, known as *backing up*, is repeated for all branches on all levels, and a single value emerges for the current position at the top; this then indicates which move to make.
>
> In this case, the left branch of the tree is better than the other two branches (whose workings are not shown). Notice that 85 is the best that can be expected because although there are boards valued as high as 440, we cannot expect to reach these values if the opponent plays well, as she will avoid these and chose a lower-valued move when it is her turn. This is known as the *minimax algorithm*; it assumes that everyone will play as well as possible, so one player will try to maximize the situation and the opponent will try to miimize it. It is a great help that chess is a zero-sum game; what is good for one player is always inversely bad for the other, and a single number can reflect the state of the game and be used to measure progress.

university project called Deep Thought, created and developed by Feng-hsiung Hsu and his colleagues in 1985 at Carnegie Mellon University.[6] In 1988, Deep Thought was the first chess computer to perform at a Grandmaster level. Then, in 1989, the inventors moved to IBM, which took over the project and renamed it Deep Blue.

When Kasparov was defeated, the popular press ran headlines like "Computer Beats Champion—Next They Take over the World," or "Chess Supremacy Spells Doom for Human Race." However, none of this has happened; the fact that a computer defeated the world champion has had no impact on other aspects of computing or applications of the technology. Chess remains a very popular game, and computing technology continues to develop apace. The triumph of Deep Blue was a great public relations exercise for IBM, but that's all it was for them. Afterward, IBM dismantled Deep Blue and did not carry on with computer chess research, but returned to its main agenda of developing and marketing ever-more-powerful processors and software systems.

COMPUTER "THINKING"

Most modern computers play chess in exactly the same way that they did in the beginning. They use the minimax backing-up search algorithm, with enhancements, specialized tuning, and efficiency innovations, but crucially, they rely on fast modern hardware to search to very deep levels and large databases for chess knowledge. However, they obviously do not play chess in the way that humans do. How could a person search through millions of moves in a second? A few researchers have looked at how humans play chess, addressing questions like: What makes a good chess player different from a beginner? Any idea that humans rely only on blind searching processes was soon dismissed; rather, people realized that the *perception* of chess positions is highly developed in chess masters. In some very early studies, both expert players and grandmasters were recorded while looking at given positions. The subjects' spoken considerations of the positions were extensively classified, processed, and analyzed, but no significant difference between the groups could be detected. And yet the grandmasters consistently found the very best moves, which the experts missed (Charness, 1981). It became clear that whatever superior expertise the grandmasters possessed, they could not explain it or make it explicit.

In another experiment, a number of chess positions were flashed up on a screen for a few seconds, and then groups of experts and novices were asked to reconstruct the position on a chessboard. Some of the positions shown were just random placements of pieces that could not possibly occur in a real chess game. The results showed that the experts performed much better than the novices in reconstructing real chess positions, but *both* performed equally badly on the nonsense positions (de Groot, 2014). Psychological studies like these show that while searching is an important part of playing chess, there is much more going on, particularly in terms of the perception and recognition of patterns and structures. Professor Fernand Gobet, an authority on chess and psychology (Gobet, 2015), has been studying many aspects of chess psychology, such as mental imagery, pattern recognition, and the playing patterns of chess players.

Unfortunately, these skills appear to be hidden in the subconscious because even chess grand-masters are unable to make their understanding explicit. Garry Kasparov has thought about this a good deal and figures that his main skills are pattern recognition—to quickly and instinctively locate the few best moves from the many possible ones—and strategic planning—that is, holding a plan in advance, which he says chess computers don't really do (Kasparov, 2017). This accords with our best understanding but, as with all other forms of expertise, despite extensive psychological research, human experts cannot fully access or articulate how their thinking actually works.

To reach master level in any particular field, there is a generally accepted rule that approximately 10,000 hours of practice is needed. Thus, if you practice for an average of 20 hours a week, it will take roughly 10 years to reach a master level of expertise (assuming you have other necessary attributes like good training sources and an appropriate aptitude). Certainly, chess enthusiasts spend many hours reading records of games, analyses, and chess media, and most chess grandmasters spend at least 10,000 hours on the perfection of their craft.

These ideas spurred a number of researchers to explore the idea of *knowledge-intensive* chess programs, as opposed to *search-intensive*. Their experimental knowledge-intensive programs were given a large database that attempted to capture both specific and general chess knowledge about positions and relationships between pieces (Wilkins, 1983). Such *knowledge bases* were loaded with all the rules, tricks, and advice on best plays in different situations that can be found in chess books and other sources of chess knowledge. Although these programs showed some promise and beat some of the very early search programs, they were soon overtaken by events, as computer hardware grew ever more powerful and enabled deeper and deeper searches.

THE OUTCOME

So, what can we learn from this condensed AI story? Now that the world's best chess player is a machine, has anything really changed? Well, despite the doom-laden headlines at the time of Kasparov's defeat by a computer, there has been little change in the chess world. Yes, we now have powerful chess machines that can beat anyone, but they also analyze chess problems, help to train players, and provide entertaining games. And people still enjoy chess and still go to chess clubs, and my local newspaper still carries a regular chess column. After his defeat, Kasparov started "Advanced Chess" tournaments, where teams consisting of a person and a computer cooperate to compete against similar teams. These teams produce high quality games, with fewer errors, and thus offer attractive spectacles for audiences to follow. Twenty years after Deep Blue, Kasparov reflected on AI and produced some insights in his book *Deep Thinking: Where Machine Intelligence Ends and Human Creativity Begins* (Kasparov, 2017). He is clearly a very competitive person and was devastated by his defeat. He has some valid complaints; for example, the Deep Blue team was allowed to change the stored chess data and evaluation function *between games*, effectively producing a new version of the program.

But apart from the saga of the chess battle, Kasparov's thoughts about AI are interesting and relevant. He describes how chess computers can find the best move but still cannot explain why it's the best. He also covers issues such as human-computer interaction, future progress, and social technology, and the book is a good read.

Modern chess programs, with names like Fritz, Rybka, Houdini, Komodo, and Stockfish, no longer need to run on a supercomputer; they now run on your home computer or laptop and have become an integral part of the chess community.[7] These widely sold products are used to analyze games and help players practice; and they are continually getting stronger. Tournaments are still held regularly, and the twenty-first century is a hive of chess activity, despite the world's best player being a computer program. In fact, chess has increased in popularity worldwide. The only people who quit chess after the Deep Blue match were the IBM computer scientists, who moved on to other challenges.

The availability of automated analysis tools has accelerated chess knowledge and understanding in humans. We now see a much higher level of skill in human games, with higher levels at amateur clubs and an increase in the number of highly rated international grandmasters. For example, in 1997, there were only 8 players rated over 2,700 on the international chess rating scale, but by 2015, there were 46, and of the

20 top rated champions of all time, only 3 played before the twenty-first century.[8] The fact that the world's best chess player is now a machine has certainly not marked the end of chess, as had been widely predicted.

What scientific knowledge have we gained from this achievement? What other areas have benefited from our increased understanding of searching techniques? Oddly enough, although they are a testament to human ingenuity, they are so specific to chess that they have little general application. IBM built special-purpose chess circuit boards for their Deep Blue supercomputer; such highly specific hardware has little use elsewhere. Even in fields where search has been a central method (e.g., in theorem proving), the tricks and techniques have been developed separately from chess programming.

Of course, there have been some amazing search developments. The truly stunning example of Web search engines is a remarkable breakthrough; to find a piece of data located worldwide, within seconds, is astonishing and would have been considered impossible only a few years ago. However, the search methods involved (e.g., the Google PageRank algorithm) have no connection with those used in computer chess. The predictions and claims made around Deep Blue proved false; the development did not translate into other areas or benefits.

In terms of Turing's views, we can reasonably say that a thinking machine for chess now exists. Turing would say that Deep Blue, or any of the current chess programs, is able to think about chess. Nevertheless, no one could sensibly claim that the thinking that these machines perform is in any way similar to human thinking in chess. A computer that thinks about chess in the same way as human players do still remains to be designed. But the next few chapters give an inkling of how this goal might be getting nearer.

OBSERVATIONS

- Intelligence is very difficult to define. AI has been seen as the automation of tasks that are generally agreed to require intelligence.
- The focus on specific tasks is similar to the situation in robotics. That is, both robots and AI systems are designed for particular tasks. Outside their task specification, though, they do not perform well.

- AI algorithms are often simple at heart; programs get complicated as modifications are added, either to deal with real-world details or to tune up efficiency to achieve maximum performance.
- When computers equal or exceed human levels of performance, media stories usually proclaim imminent doom and disaster. But the opposite often happens. The defeat of Garry Kasparov by a computer did not mark the end of the game of chess. In fact, chess activity actually grew worldwide, increasing human chess knowledge and producing much higher levels of human skill. Chess has never been so much fun or so popular.
- Although search is still a powerful technique, research strives to reduce dependence on brute force methods (as will be seen shortly).
- AI solutions are usually not based on methods that humans might use to solve a problem. This is often provably or self-evidently true, as in the case of deep searching.
- Human expertise takes a long time to acquire and is often encoded unconsciously as covert knowledge, which is inaccessible even to the owner.
- Predictions of future AI performance are usually wrong, even when given by experts. Predictions on the implications of technological breakthroughs are just as bad.

5

KNOWLEDGE VERSUS POWER

Thoughts on Knowledge
- Data are symbols or signals.
- Information is the detection or extraction of meaning from data.
- Knowledge is the understanding of information.
- Wisdom is the reflective use of knowledge.

Searching is a relatively straightforward process; knowledge is a bit more difficult to understand. As stated previously, intelligence is hard to define, and it is similarly difficult to pin down just what aspects of knowledge are important for computer use. Put another way, what do artificial intelligence (AI) programs actually *know* about the things they are processing, about the domain they are working in? This is actually a very important issue in AI, and it is also crucial for robotics.

Consider those chess programs again. There is an important trade-off in these game-playing programs, which also applies to many other search-based systems. The trade-off is between depth of search and the degree of knowledge required. If you have a lot of knowledge about the game, then you don't need to search very much at all. On the other hand, if you have no knowledge (other than the rules of play), then you must search very deeply down the game's search tree to find good moves to make.

To appreciate this key point, we need to look at the way that chessboard positions are evaluated. When the generator reaches a certain depth (as set by time and memory limits), it calls the evaluator to make a judgment about the board position at that point. How can chessboard configurations be given values?

To figure this out, we have to use basic knowledge of the game. Several features of a good board position are calculated separately and then combined into one overall value. One key feature is material advantage. For example, a queen is considered equal to a certain number of pawns (usually 9), so a side that loses its queen is (at least roughly) worse off than the opponent who is 4 pawns down. By counting up the pieces on each side and assigning higher values to the major pieces, it is possible to give a single number to represent material advantage. If both sides had exactly the same number and types of pieces, then the advantage would be zero, but for more or better pieces on the side of Black (or White), the value would be positive (or negative) and would increase with the strength of the advantage. Other features that are often calculated include the number of available moves (piece mobility), the degree of control of the center, and the number of possible captures. These game-specific features are combined in various proportions in an equation to achieve a final board evaluation. Thus, a wide range of specialized considerations are condensed into a single numeric value. These features apply only to the game of chess—any other game will require its own special, handcrafted, evaluation function.

We can now see that knowledge of the game is contained only in the evaluator; the generator simply carries out a mechanical hunt for the best value it can find for the next move. Now let's take this to extremes. If we could search to an unlimited depth, then the evaluation function would be extremely simple: just give a positive or negative value, depending on who won (or zero for games that end in a draw). The search process would terminate only on positions that were at the end of the game (win, lose, or draw), and so no real game knowledge is needed. At the other extreme, assume that we have been very smart and produced an evaluation function that contains an awful lot of high-quality, distilled chess knowledge.

In this case, the evaluation function might be so good at judging board positions that we could simply let it look at the current chess board to assess all the available immediate next moves and take the best one. No search would be needed at all. This approach has some advantages because the savings in processor power through not searching can be applied in computing very complex evaluations. Such knowledge-intensive methods have been tried, but we currently don't have a solution to the problem of extracting the important covert knowledge and getting it into a written form. Compared with this, automatic search seems easy!

The Nobel Prize–winning psychologist Daniel Kahneman describes human thinking as consisting of two systems: one for rapid intuition and one for logical, considered reasoning. Kahneman describes them thus: "System 1 operates automatically and quickly with little or no effort and no sense of voluntary control. System 2 allocates attention to the effortful mental activities that demand it, including complex computations. The operations of System 2 are often associated with the subjective experience of agency, choice, and concentration" (Kahneman, 2012, 20–21).[1] This distinction resonates with our everyday experiences of thinking. On one level, we use intuition, remember things, instantly recognize a friend, and other actions, all without any conscious thinking; they just happen. On another level, we make decisions, compare options, and ruminate on a problem, in what seems like a deeply personal, effortful thinking process. In other words, one is very fast and not available to conscious thought, while the other is slower, requires conscious attention, and is subjective.

These hidden assumptions and intuitions create big problems with automating apparently simple tasks. As soon as we try to implement an artificial intelligence (AI) solution, we require knowledge that is usually glossed over in human thought. Looking again at Gwen's trip-planning exercise in chapter 4 illustrates this clearly. Recall that Gwen wanted to go from Cambridge, England, to Fort William, in Scotland. In figure 4.1, Gwen immediately ruled out flying—but why? She could get to Glasgow Airport from Cambridge, and then to Fort William from Glasgow Airport by bus or train, and a computer search would have produced the results shown in figure 5.1. You can't rule out the nearest airport just because it is

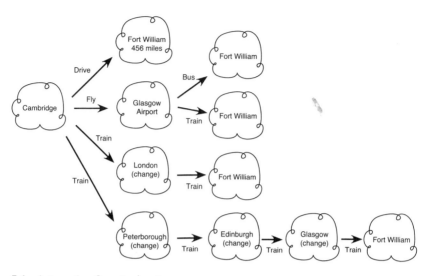

5.1 Automating Gwen's planning.

not the final destination. So she must have known that Glasgow is too far away and the bus and train journeys from there to Fort William are too long and tiring (in her opinion). This knowledge was directing her search, and it would be required in a similar computer planning assistant. Also, Gwen could have gone by train on a more direct route, avoiding London and getting there more quickly. However, this requires the use of four trains, and she probably knew that fact but wanted to trade comfort and convenience for time—another bit of knowledge the computer would need. Without such guiding information, we get brute force searches, which terminate only upon achieving their goal, as seen in figure 5.1.

These issues—subconscious preferences, conscious thoughts, assumptions, and bias—all represent different kinds of knowledge. Box 5.1 shows different ways in which we may "know" something.

HOW CAN KNOWLEDGE BE STORED FOR UTILIZATION?

A closely related, but more fundamental problem concerns how computers can best store and process knowledge. Data is everywhere and consists of signals or symbols, but they only have significance if someone can interpret the information they contain.[2] For example, a message

Box 5.1
What is knowledge?

> When we ask, "What does this person (or system) know?" it is important to recognize that knowledge can be held in many forms.
>
> **Body skills** are programmed or learned within the musculature and motor components. These are noncognitive but are knowledge nonetheless. We all have to learn how to ride a bike, even though we can't mentally determine if we have this ability or not.
>
> **Facts** are statements that are either true or false (in terms of pure logic). But in human cognition, statements can be held that may be true, could be in conflict with other facts, or are known to be temporary. Thus, humans have a rich factual base that includes beliefs and facts of dubious justification.
>
> **Episodic memory** stores knowledge of sequential events or procedures. Recipes and maintenance manuals are examples where the sequence and timing of events are important.
>
> **Rules** are relationships that hold between items of knowledge. They are a kind of fact that can capture definite relations, such as connections between other facts, or more irregular relations, such as rules of thumb, known as *heuristics*.
>
> **Memories** include stored knowledge in any form. This covers all of these points and includes items such as unattached images, events, and experiences that as yet have no relation to anything else.

in a foreign language is clearly data, but it carries no information if you don't know that language. Knowledge can be seen as an assimilation of information; information may convey meaning that increases the knowledge of the recipient. During the 1980s, AI researchers were fascinated with designing and building large knowledge bases. The knowledge that experts could build up through extensive experience seemed to offer a promising analogy for intelligent systems that no amount of brute-force computing was likely to achieve. These knowledge bases are not simply large databases of facts; they are systems that store vast quantities of data in a suitably structured form that can capture information and knowledge in a way that humans can relate to and use. One of the necessary tools for knowledge processing is an *ontology*, which is a kind of highly structured dictionary that records the essential properties of things. The term *things* here means concrete items, like objects; abstract concepts, like ideas; and relations, like effects and influences. If one has a good ontology for a particular problem domain, then it becomes possible to act meaningfully in that domain. An ontology acts like a kind of reference book that can be consulted to find out the basics of any topic. Then the stored facts in the knowledge base can be exploited and reasoned about during interactions or other applications.

Various languages have evolved to help with knowledge systems, the idea being to capture the structure of the data in a kind of description. We may be familiar with Hypertext Markup Language (HTML), the markup language used by web browsers that describes the presentation of material on web pages. A series of these *metadata* languages has evolved, producing increasingly advanced markup systems that try to represent not only presentation and structure, but also meaning.[3] This leads into knowledge representation systems (with names like KL-ONE) that are formal systems, based on mathematical logic to ensure that they have predictable properties. The main idea is to provide some semantics, or meaning, for knowledge bases so that retrieval applications are not just syntactical, but also have a deeper understanding of the content. This is particularly relevant for the World Wide Web—the next generation of the web has been called the *Semantic Web*, and the name promises richer, deeper interactions with the web when we search for data or interact with applications. However, implementing the Semantic Web involves a great deal of international effort, including standardization and engineering work. This has turned out to be a significant obstacle, and it has been a long time developing.

COMMON SENSE KNOWLEDGE

A classic AI knowledge project currently under way is attempting to create an encyclopedia of every-day, common sense reasoning and general knowledge. This project, called Cyc, is building a comprehensive ontology and a knowledge base. Its originator, Doug Lenat, argues that this is necessary in order for AI applications to perform humanlike reasoning. Millions of pieces of common sense knowledge are being encoded. This is the sort of knowledge that is so obvious that it has never been formalized before (except in some scientific work). It includes concepts like gravity, the uniqueness of individual entities, and the persistence of objects. Much of this is known as the *physics of the world*; i.e., the way that simple objects behave under gravity, friction, inertia, motion, and other forces. Other such "common sense" knowledge, which we never really notice but use all the time, includes universal principles like "an object can only be in one place at a time."

For example, one of the easiest things for a computer to do is clone (i.e., make perfect copies of) things. Given a file, picture, or folder, a computer can create a copy instantly. Your personal details that you have entered into a few web-based services now probably have hundreds or even thousands of copies floating around on the internet. Such clones do not make sense in the human world (the real world!); every physical entity is a unique individual, and even biological clones immediately diverge and change according to their own life histories. But any computer that is going to "understand" the human and physical world has to be given a lot of extra rules just to prevent mistakes like deciding that "Mary is cooking in the kitchen" and "Mary is asleep in the bedroom" are both true at the same time. This type of knowledge error seems obvious and trivial to us, but it can be difficult to model and capture in knowledge-based systems.

Shortly after the project started in 1984, Lenat estimated that the effort to complete Cyc would require 250,000 rules and 350 human-years of effort. It now has over half a million concepts in its ontology and five million facts, rules, and relations, and the project is still ongoing. The knowledge and rules are created by a team of professionals known as "knowledge engineers," who collect and enter facts about the world by hand and also build ontologies and implement inference mechanisms for the knowledge base. An open-source ontology system called OpenCyc (version 4.0, 2012) can be browsed (see www.cyc.com), giving access to hundreds of thousands of terms, along with millions of assertions relating the terms to each other. Other available components include the Cyc Inference Engine, natural-language parsers, and English-language-generation functions.

The reason Cyc is such a huge task is because there is no way to automate the construction of what is, essentially, an "ontology of everything." Encoding all that humans know, both explicitly and implicitly, is a vast task, and Cyc is a bold undertaking. It has been suggested that Cyc could help in the creation of the Semantic Web by using its huge knowledge base to provide an account of the meaning of the information on all web pages. Unfortunately, the World Wide Web presents two additional problems for Cyc: It is several orders of magnitude larger, with around 5 billion pages; and it is a more malign environment. Not only is it full of uncertain, inconsistent, and conflicting information, but some of it is also deliberately deceitful and destructive.

SEARCH IS A STANDARD TECHNIQUE

Before we leave the topic of search, it is worth pointing out how important it is in AI and computer science (search is ubiquitous in life too–we all perform searches every day). Search is simple in concept but can be very powerful. When you vaguely remember that there is a document on your laptop, you can find it by typing a distinctive word that it may contain—it will take just a second to run a search that examines *all* the words in *all* the documents on your computer and find all the documents that contain the search term.[4]

AI has produced many variations and advances on the basic search idea. Pruning techniques reduce the size of a search tree by preventing some branches from being examined. A method called *alpha-beta pruning* can reduce a tree search by 60 percent (by not following branches that can be seen to be worse than those already explored). If the costs of examining branches can be estimated, then these costs can be incorporated into the process, and a kind of shortest-path estimate helps to increase efficiency—the A* method (known colloquially as "A star") does this and is widely used in computing and AI.

In addition to planning, search is the bedrock of reasoning and formal problem solving. We perform reasoning when we diagnose a problem, solve a puzzle, or resolve a set of conflicting demands. Similar to planning, reasoning involves looking at options, weighing them, and deciding which path is the best bet. AI research on reasoning has extended from the purely logical to include more human-centric ways of reasoning, such as fuzzy, qualitative, and analogical reasoning.

Formal problems are those that have strict rules and goals. Automated theorem proving is the classic example of problem solving in computer science. It consists of finding a series of steps that logically lead from a group of given statements (all assumed to be true) to another statement (the *conclusion* or *theorem*). This is a rather theoretical area, but it forms a fundamental part of computer science and the theory of computation.[5] It has also produced valuable applications, such as verification techniques for proving the correctness of programs. This is of great importance in safety-critical situations where software reliability is vital. For an introduction to search (and indeed all mainstream AI techniques), a good source is Russell and Norvig (2010).

SYMBOLS AND NUMBERS

There are two fundamentally different ways of representing knowledge in AI: textually and numerically. Practically all human knowledge is recorded in language, usually in the form of words and text. Images and graphics are also important, but they are usually associated with text. When information is coded in the form of alphabetic symbols, the inner workings of AI systems that process such knowledge can be examined, read, and understood by humans. For a long while, AI was focused on symbolic processing; it seemed to be effective and was thought to resemble human processing. Mathematical logic even became a standard formalism for implementing many theories, including natural-language processing and acquisition. Humans are clearly able to process symbols, but as Hutchins (1995) says, they may process internal representations of symbols, but this does not mean that symbol manipulation is the architecture of cognition.

However, when neural models came on the scene, all the information was processed quite differently—namely, by being encoded and distributed across a large number of variable numeric parameters. Consequently, the knowledge that these systems gained was impenetrable and difficult to interpret. This created the distinction of a class of *subsymbolic* AI systems. To some extent, the new statistical methods encode their knowledge into numbers as well, although they can handle text as a by-product.

LEARNING TO IMPROVE

A mixture of knowledge and search makes a good compromise—learn as much as you can (knowledge), and then apply it to plan through the various options (search). We know that chess masters do this; they recognize certain scenarios and then do mini-searches to test out the options. But the basic computer game-playing methods have one severe limitation—they do not include learning. If you play exactly the same game with them a second time, they will lose or win in exactly the same way. Thus, the next development was to implement some form of learning to enable performance to improve with experience. An important pioneer in this area was Arthur Samuel, who wrote two seminal papers on computer learning for playing checkers (Samuel, 1959, 1967). Checkers, known as *draughts* in the UK, is a board game played on a chessboard with 12 dark and 12 light pieces. It is not an easy game, but it is considerably easier than chess.

The complete search tree for the game of checkers has now been fully explored. An enormous amount of computation, across hundreds of computers, resulted in a proof that, providing both players play as well as they possibly can, the game always ends in a draw. This result, in 2007, followed all the perfect lines of play and processed 10^{14} board positions (Schaeffer et al., 2007). That's 100,000 billion board positions. Note that they only covered the best moves; they didn't generate the search tree for all the other moves—that would require 500 billion billion possible positions (5×10^{20}) to be examined. But even if we might manage that with today's supercomputers, there is little chance of enumerating the tree for chess, which has about 10^{45} positions (i.e., 1 with 45 following zeros!).

Samuel realized that the mixture of features in the evaluation function could vary. For example, a given function might calculate the value of a position as follows: 60 percent of material advantage, plus 30 percent of control of center, plus 10 percent of capture possibilities. But this mixture, as determined by the numbers 60, 30, 10, is set by the designer and is simply an estimate of what might be important: No one knows what the best mixture should be! Instead of fixed proportions for the mixture, Samuel let them be parameters, x, y, z, and arranged an adjustment scheme so that they could be altered according to game performance.

His innovation was to pit two programs against one another, playing many games of checkers. One program would have its parameters held fixed for the game, while the other could adjust its parameters while it learned. When the learning program performed better than the fixed version, they both would be given the new parameters and then play again. The idea was that the evaluation function would only improve, and so they would bootstrap each other onto better and better evaluations, and thus to better playing skills.

There is an apocryphal story that Samuel spent some time as a nightwatchman at IBM, looking after a huge warehouse full of IBM 360 computers, and he had them all playing checkers with each other all night long. (One computer can play a full game by switching between the two opponents. Interestingly, this is something that humans find very hard or even impossible to do—that is, within a single brain!)

Samuel also pioneered many of the little programming tricks and techniques that became widely used in systems playing board games, and his papers remain very readable and relevant for students of computer games. But after 10 years' work, he was still worried that he might not have found the very best features and couldn't find a way of "getting the program to generate its own" (Samuel, 1967, 617). He had automated almost everything he could, including learning the feature parameters, but he still had to design the actual features themselves. He saw that using the right features is crucial for good performance, and what was really needed was a method for the computer to discover the best and most appropriate features itself.

In his second paper, Samuel says that automatically generating the features "remains as far in the future as it seemed to be in 1959" (Samuel, 1967, 617). His technique to deal with this problem was to think of as many features as possible, implement them all, and hope that the most effective ones were included. He usually experimented with around 25 features, but often tried as many as 40. During learning, some features would drop out as their parameters approached zero; they were bad features, with nothing to contribute. But how do you know if you have missed a *good* feature? And what is a good feature anyway?

FEATURE ENGINEERING

The question of what is a good feature is a major problem that has dogged many branches of AI for decades. Consider the case of visual classification. Suppose that we wish to automate the sorting of a series of flowers of different species. We want to build a sorting program that will learn the different categories of flower by seeing images of typical examples of each one. The usual way to approach this is first to choose a set of distinguishing features; these could be any visual properties of the images that might be useful in recognizing the flowers, such as color, shape, texture, and particular patterns. Second, we train the system by presenting a set of images containing many examples of each category to be learned. The learning program adjusts its parameters as each example is seen and eventually settles on a description of each category in terms of a particular mix of features. So class A might be defined as x amount of feature 1 (leaf shape), y amount of feature 2 (predominant flower color), and z amount of feature 3 (head shape). Classes B, C, and so on will have distributions of features represented by sets of different values for x, y, z. Finally, when the program has seen sufficient examples, it can be used to recognize new, previously unseen flowers by matching the feature content of a new image to the category with the nearest mix of features.

The key process here (for visual recognition)—classifying by feature—is exactly the same as for the evaluation function in game playing, and also similar to many other computer learning domains. The key issue of the origin of the features is thus common across many problems and methods. We can design systems that will learn all kinds of patterns, but the actual features used, which are really primitive fragments of knowledge about the application and which are so crucial, are essentially educated guesses handcrafted by designers. While learning and recognition systems have achieved good results over the years, generally improving the methods and building on application experience, the results depend on engineering aspects like feature recognition to suit the applications. The lack of a scientific method for finding good features has been a persistent theoretical and practical weakness of AI—that is, until recently!

OBSERVATIONS

- Search is a well-developed technique, and it is used in many areas of computing, including formal methods, logic and language, and planning and reasoning.

- A little knowledge can save much brute-force exploration of a search tree.
- Deep searching on poor-quality data can sometimes produce good results. Conversely, a rich supply of information can remove the need to search deeply, if at all.
- This important trade-off between deep but mechanical searching and available knowledge offers opportunities to find different solutions to AI problems.
- The capture and processing of knowledge are poorly understood, mainly because they represent some of the most obscure aspects of human cognition.
- Human knowledge involves much implicit, experiential information that cannot be made explicit, and knowledge bases that attempt to capture such knowledge have been handcrafted, involving enormous amounts of effort.
- There are many different kinds of knowledge: factual data, episodic data, sensory-motor knowledge, relational knowledge, and all kinds of images, sounds, and patterns.
- Ontologies are popular for encoding kinds of knowledge, structuring knowledge bases, and assigning meaning, but they are built and maintained by humans.
- The so-called higher cognitive functions, such as language, reasoning, and planning, traditionally have used symbolic AI techniques, especially logic-based methods. The more subconscious aspects of intelligence, such as vision and speech, more often have been handled by subsymbolic AI techniques usually based on artificial neural networks (ANNs).
- Handcrafted features have been a long-standing weakness in AI. The ideal is to achieve fully automated systems without any prestructured features.
- The recent enormous increases in computer power have allowed some old, sometimes simple, and previously impractical methods to become feasible and useful. We can now search through reams of data in seconds.

6

A LITTLE VISION AND A MAJOR BREAKTHROUGH

Any sufficiently advanced technology is indistinguishable from magic.
—Arthur C. Clarke's third law, in *Profiles of the Future*, 1973

The branch of artificial intelligence (AI) known as Deep Learning has been responsible for much media excitement, as well as often overhyped predictions. Deep Learning has had spectacular success in vision and imaging applications, speech and language processing, pattern and data analytics, and game-playing. A short explanation of computer vision processing will set the scene for this major development.

It has long been known that visual processing in the brain involves building up complex patterns from hierarchies of very simple visual fragments or motifs. The fragments are analogous to the features we discussed in the previous chapters. In 1981, the Nobel Prize for Physiology or Medicine was awarded to David Hubel and Torsten Wiesel for their work on the visual cortex. In a seminal paper, Hubel and Wiesel (1962) reported on the organization and function of the brain cells in the very early stages of the visual cortex. They showed how input from the eyes enters an area of the brain known as the *primary visual cortex*, or V1, where the brain cells of the nervous system (neurons) are arranged as an accurate topographical map of visual input from the retina. Individual cells detect either color values or the orientation of contrast changes in the image. Visual area V2, or the *secondary visual cortex*, is similar to V1 but can detect various shape characteristics. The outputs from these layers are sent to the areas V3 (which is not color sensitive, but detects dynamic movement of shapes), V4 (which processes color and shape), and V5 (which is interested only in movement and is sensitive to different directions of movement). These layers of feature detectors are organized into two streams. One stream is concerned with the perception and recognition of objects (involving V1, V2, and V4) and is sensitive to color, shape, and orientation.

At each stage upward through the hierarchy, the processing produces more refined and complex outputs that focus on specific information, so this stream eventually reaches brain areas where the cells can detect complex patterns, such as a particular face. The system generates a response to the face by effectively assembling lots of little lines and patches of color into motifs such as eye components, head features, and nose and mouth fragments, and then further combines them until a layer is reached where they are combined into recognizable entities that can be compared with previous memories. The other stream (involving V1, V3, and V5) largely ignores features like color and shape because most of the cells are tuned to the speed and direction of moving visual stimuli. This is valuable for dealing with the position and motion of objects in space, self-movement, or environmental changes. These two streams are often informally called the "what" and the "where" pathways in the visual brain.[1]

The surprising thing about the feature detectors in the brain is that they are so basic, so primitive. For example, figure 6.1 shows a simplified arrangement from a cell in V1. The X and O symbols indicate topographic

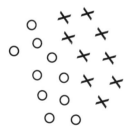

6.1 An arrangement of inputs for an edge detector cell. If the pixels marked O are light and the pixels marked X are dark, then a contrast edge has been detected (at a not-quite-vertical angle).

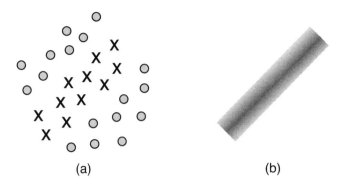

6.2 (a) Inputs for a line detector cell; (b) a section of line image that would match the detector.

inputs to the cell from the retina; that is, they are in the same spatial arrangement as the eye receives the image. This cell fires (i.e., gives an output) when most of the Xs are active and most of the Os are inactive—this corresponds to high and low image intensity (i.e., black and white)—and so this cell responds to a high contrast change along an approximately straight line. This is known as an *edge detector* or *edge filter*. Note that this design is orientation sensitive; if the angle of the visual edge changes (say, to horizontal), then the pixels will not match the detector cell and there will be no output.

Hubel and Wiesel reported that such cells are sensitive (or tuned) to within 5 to 10 degrees of their preferred orientation. So in order to sense edges of different angles, the brain must have a large population of cells that detect edges for a wide range of orientations. Figure 6.2(a) shows another type of V1 cell; the arrangement here is sensitive to a parallel, double edge; that is, it is a *line detector*. An example image element that would trigger this line detector is shown in figure 6.2(b). These line or edge detectors have specific angles of orientation, but they will also work if the edge is a little bit rough; not absolutely all the inputs have to be correct, and this allows a degree of tolerance.

Computer vision-processing systems use the same idea: Figure 6.3 shows a typical 5 by 5 pixel detector that could be used in a visual recognition program to find lines of approximately 45 degrees in an image. This detector, or *convolution filter*, has been designed to give an output when the 5 marked cells are active and the other 20 are inactive (we don't

6.3 A detector for lines at 45 degrees.

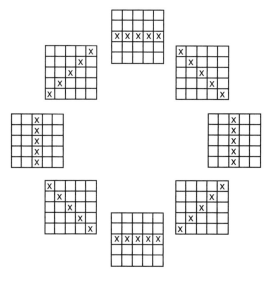

6.4 A set of line detectors (of four types) arranged spatially to locate a roughly circular line.

need to show the Os). These filters are much smaller than the image (only 25 pixels, in this case), and the whole image needs to be processed by trying out the filter over the entire image. This can be achieved either in parallel (using special parallel hardware) or by sequential scanning of the filter (in software).

By combining a range of these primitive filters, we can detect more complex and interesting patterns. Figure 6.4 shows how eight line filters can cover parts of a larger line pattern. In this case, if they all give an output, then an approximation to a circular or closed form is being signaled. The circle detector is made of only four line orientations: horizontal, h; vertical, v; diagonal 45 degrees, d; and diagonal 135 degrees, b. These can be combined to make a single circle detector, and figure 6.5 shows how this is done in a hierarchical scheme.

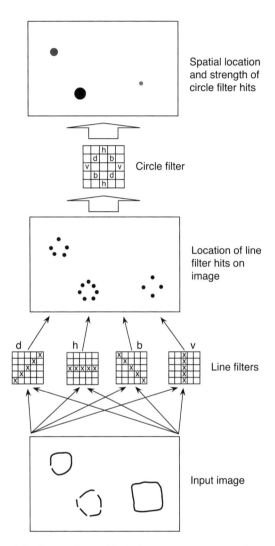

6.5 Three hierarchical layers, processing from images to line fragments to shape detection.

The raw input image is scanned by all four line filters, and their outputs are stored in another image-sized layer. This layer indicates the locations where line fragments have been detected; these are shown as dots indicating filter "hits," but they will be coded with the orientation of the filter. Next, a single "shape" filter is scanned over all locations in the line layer, and its output goes into the next higher layer. Clearly, the shape filter will give strong output when all eight of its specified cells match the desired orientations, and this is best for circular shapes. But less perfect shapes will also give some response—the square-ish closed shape in the lower right quadrant of the image gives some filter hits, and so a weaker output in the top level indicates that some approximately circular form has been seen. These layers can continue until, ultimately, they reach cells that can detect faces, cars, and many other objects. The topographic relationships with the image are maintained by the layers, although the size of the higher layers often is reduced.

This is a basic scheme for computer-based vision processing and pattern recognition, inspired by what we know of brain science. Many kinds of classification or learning methods have been built on top of this image-processing technique, but the basic method of hierarchical layers of successive fragment extraction has been used for the last 40 years. All kinds of image-processing applications have been developed; medical (e.g., X-ray and scanner analysis), industrial (e.g., process inspection and robot control), monitoring and tracking (e.g., traffic and pedestrians), security (e.g., face recognition)—in fact, any area where images are involved can use image processing to extract useful information. But these are programs designed by AI researchers and software engineers and, as for games, the feature detectors have to be designed as well. That is, until now!

THE END OF FEATURE ENGINEERING

Just a few years ago, a real breakthrough was achieved. Krizhevsky, Sutskever, Ilya, and Hinton (2012) reported on some spectacular results obtained in a visual recognition challenge called the ImageNet Large-Scale Visual Recognition Challenge (LSVRC).[2] A team from the University of Toronto, led by Geoffrey Hinton, almost halved the best previous error

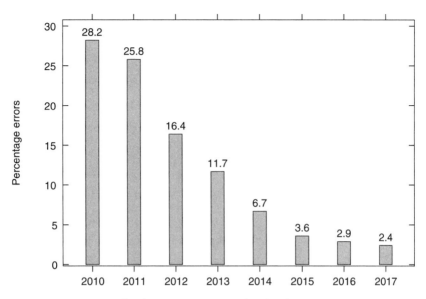

6.6 Percentage errors for the winning entries in the LSVRC competition.

rate for object recognition. This was a major advance because such error rates usually reduce very slowly as the methods are gradually refined and improved. This drop (in 2012) and subsequent progress are shown in figure 6.6. The Toronto team trained a very large artificial neural network (ANN) on 1.2 million color images from a standard data set in order to recognize 1,000 different classes.[3] Their network had eight layers, with 650,000 artificial brain cells, and had to learn 60 million parameters. To maximize speed, the effort was implemented on special hardware chips developed for real-time visual gaming systems [known as Graphical Processing Units (GPUs)], but even then, it took more than five days to train using two GPUs.

This was clearly not a freakish blip, but rather a real breakthrough because the same approach has been used in all subsequent competitions and the error rate continues to drop dramatically.

There has been a 10-fold reduction in image classification error since the challenge competition started in 2010—from 28.2 percent in 2010 to 2.4 percent in 2017. A clearer way of showing the performance improvement is to chart the reduction in error as a proportion of the previous best results; figure 6.7 shows this reduction as a percentage. In 2011,

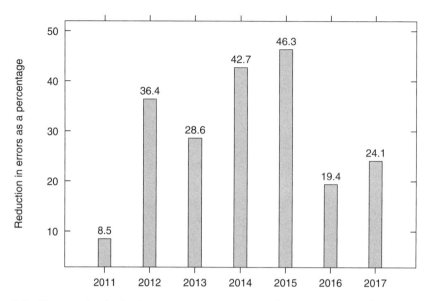

6.7 The annual reduction in errors as a percentage on the previous year in the LSVRC competition.

the improvement was very typical of the gains produced for many years–around 10 percent at best. Then, in 2012, we saw the dramatic error reduction of 36 percent from the first Deep Learning experiments. Although the improvements don't look too astounding in figure 6.6, when shown as in figure 6.7, we see that, even with low errors, in 2015 the reduction was nearly 50 percent! This is remarkable and reflects the qualitative change of technique being employed. The error rates are now so low that there is little room for improvement. Performance is now approximating human levels on these tasks.

This advance stimulated much interest in deep networks, as can be seen in the dramatic participation increase in the challenge, from 24 teams submitting in 2013 to 84 teams in 2016.

This is a qualitative breakthrough, not just a quantitative improvement, because the machine can now *learn for itself*. This has solved the problem of hand-crafted features, as these can now be discovered automatically. By being learned from large numbers of examples of real data, they can be tuned for the recognition task itself. These visual processing networks are often called *convolutional neural networks* (*CNNs*), or

ConvNets, and the general approach is called *Deep Learning*, with the word *deep* here referring to the number of layers used in the neural network.[4] Interestingly, when people examine one of these deep image-learning systems after they have been well trained, they find that the first layers have learned filters that look very much like the edge detectors and other functions seen in V1. Perhaps this is not surprising—if evolution has found them so useful, then they should be well suited to the job.

Another impressive example of this technology is the deep network that learned to play 49 different video games on the Atari game console. The Atari 2600 was a popular arcade-game computer from the 1980s that ran games like Pac-Man, Space Invaders, Breakout, and Pong. They were all played on a video screen of 160 by 210 pixels, with 128 colors. (The coarse blocks of garish colors remind us how far video processing has progressed!) The player had a joystick that could be moved in eight compass directions and a single button; this gave a total of 18 states that the player could select as actions. This rather limited input and output means that the games are not very complex but still pose a significant learning challenge. A software simulator for Atari had been developed, which was used to experiment with a convolutional network of five layers, with the inputs being the raw image data from the Atari screen and the current game score, and the output being a signal indicating 1 of the 18 possible states for the joystick and button. The network (known as DQN) used a form of *reinforcement learning*, wherein the score represents a reward and the system tries to find sequences of action that maximize that reward. The idea was that the system had to face the same problem as a human when first learning an Atari game—the only inputs are the visual images from the screen, and the only actions occur via the joystick. After training, the equivalent of 38 days of game playing, there were some striking results. In nearly all the games, DQN was better than other automated learning methods, and for 29 of the 49 games, DQN was as good as, or superior to, a professional game tester.[5] The key to this success lay in using the popular method of reinforcement learning to assign the credit for a win to the critical action or actions, which may be several steps before the actual winning move.

These Atari experiments were carried out by DeepMind Technologies, a British start-up founded by several academic game developers. DeepMind was bought by Google in 2014 and is now known as Google DeepMind. Most of the pioneering experts in the field of Deep Learning have been acquired, in one way or another, by the larger companies like Facebook, Microsoft and IBM, where they have easier access to large resources compared with some of their academic environments.

To appreciate the deep approach in a completely different environment, we can return to two-person board games, in particular the Chinese game of Go. The game of Go is extremely difficult—much harder than even chess. It is a zero-sum game like chess, but is played on a 19 by 19 board with black and white pieces. The increase in locations on the board, from 64 to 361, accounts for much of the extra complexity. The number of possible board positions for the black and white pieces is much higher than for chess: 10^{170}, or 1 with 170 zeros (an *enormous* number; try saying "billion" 18 times). A team of experimenters from Google and Toronto University built a CNN to learn an evaluation function for Go (just like the one described for chess) (Maddison, Huang, Sutskever, and Silver 2014). The network was large, with 12 layers containing more than a half-million artificial brain cells and 630 million connections. The learning process involved tuning 2.3 million parameters. They trained this CNN on a database of 27 million expert moves, and then it was tested by showing it 2 million board positions and noting its best move outputs. The results gave a 55 percent prediction accuracy for the best move, which is very good when compared to the previous best of 39 percent by other methods. The authors then tried using the CNN to play the game, but without any searches, just by choosing the best-looking move for the current board position. The CNN equaled the performance of the best search programs and beat a standard search program (GnuGo) in 97 percent of the games played.

Recently, the program finally beat the world champion Go player! In October 2015, Google DeepMind announced it had incorporated its Go evaluation function into a search program called AlphaGo, and it beat the European Go champion, Fan Hui, 5 games to 0. And then, only five months after that, the program played the world champion, South Korea's Lee Se-dol, in a five-game match and won 4–1. During that five-month gap, it self-improved by playing against different versions of itself millions and millions of times.[6]

It was clear that with such good results from their CNN evaluator, a large amount of quality Go knowledge had been captured—and it could play well without *any* searches. The depth-versus-knowledge trade-off means that adding even a small amount of search functionality would increase performance dramatically, which it did.

The real surprise is the speed that these advances are being made. One author, reviewing the level of game-playing AI programs in 2014, stated that Go programs were at a "very strong amateur level" and "might beat the human world champion in about a decade" (Bostrom, 2014). It is a measure of the rate of progress in Deep Learning that it took only 20 percent of that predicted time. Further breakthroughs in this story are covered in the next chapter.

WHAT HAPPENED?

Why are these Deep Learning experiments proving so successful? It's not really because of any big theoretical breakthroughs. The learning techniques used in the Deep Learning networks are not new; rather, they are tried and tested methods like backpropagation, reinforcement, and gradient-based algorithms, which have been around for at least two decades. Indeed, successful versions have been implemented since the 1990s for reading handwritten digits on bank checks and ZIP codes.[7] A digit recognition system using this approach has been reading millions of checks in banks across the United States since 1996 (LeCun, Bottou, Bengio, and Haffner, 1998). So the technology was not the problem.

Most previous research always used small networks, often with only three or four layers, because it was widely believed that more layers would not bring any real benefits. This seemed to be confirmed in practice because when scaled-up designs were attempted, the results usually worsened. In addition, there simply wasn't enough available computer power at that time to experiment on larger networks. Therefore, researchers became discouraged from working on larger networks—training so many variables looked impossible.

It seems to be a case of "size does matter." The two notable characteristics in all these Deep Learning experiments are that the networks are huge, with millions of elements and billions of connections, and the training data set is also huge, drawing millions of examples from gigantic databases. These size increases have been possible recently only because of the massive computer power that is now readily available and the supply of very large data sets on professional servers supported by global organizations. The size factor effect has been reinforced by work that suggests that the large number of layers is really necessary. In the visual recognition challenge competition, the Toronto team, Krizhevsky et al. (2012), tried removing layers from their eight-layer system, but any single layer removed resulted in a drop of performance of around 2 percent. So depth does seem to matter; small networks with a few layers can be made to work on small problems, but really big problems need really big networks.

Deep Learning is here to stay. Some spectacular results, not to mention the excellent performance on solving difficult problems, have caused many major companies to look into developing applications for all kinds of new technologies. Deep Learning systems are being developed into cheap, efficient packages based on specialized hardware for vision-processing applications in self-driving cars, phones, video systems, and every other area of consumer technology. The field of image processing has been profoundly affected by this breakthrough, and Deep Learning methods are now overtaking traditional vision-processing methods. Computer vision conferences and challenge competitions regularly report that the majority of the submissions use Deep Learning, and methods involving this usually outperform all the others. They are not perfect, but everyone seems to be using them.

OBSERVATIONS

- Image-processing software has a long history of using hierarchical feature detection, with some similarities to how the brain works. These layered hierarchies are a natural fit for implementing in ANNs.
- Feature engineering, an old bottleneck, has finally been overcome. Deep Learning of application-sensitive features can now replace the handcrafting of features in image systems, games, and many other AI areas.
- Deep Learning has had tremendous success, breaking records in vision and speech performance, and games like chess and Go.

- Deep Learning involves very large, multilayered neural networks (typically with millions of parameters to be learned). They are frequently trained on enormous amounts of training data (typically millions of megabytes).
- Increases in the performance of hardware and software tools has allowed anyone to implement Deep Learning methods, either by buying special hardware or by renting powerful resources on the cloud.
- Breakthroughs are not always entirely original or about new knowledge; instead, they can be based on new ways of doing old things. Deep Learning is really *deeper* learning, but the qualitative improvements are enormous.
- These new learning methods, including techniques like reinforcement learning, look as though they soon will be almost ubiquitous in AI. But AI is a dynamic field that can change direction very quickly.

7

THE RISE OF THE LEARNING MACHINES

We can be knowledgeable with other men's knowledge, but we cannot be wise with other men's wisdom.
—Michel de Montaigne, *Essais, Book I,* ch. 25, 1595

One of the strange things about the early decades of artificial intelligence (AI) was that learning was largely ignored. Researchers were writing programs that could recognize visual scenes, perform logical reasoning, or plan a complex task, but learning was not a big part of these systems. The dominant view seemed to be that AI was a kind of puzzle, or set of puzzles, that might be cracked by finding the correct solution or solutions. And the different dimensions of intelligence could be dealt with through separate projects for each puzzle. Hence, AI was split into various streams, and a typical list of AI topics would cite automated reasoning, problem solving, knowledge representation, planning, learning, natural-language processing, visual perception, and robotics. Learning was included as a problem area, but it was usually treated as a generally desirable property of AI systems rather than as a topic on a par with the other subjects of study.

So the computer systems created by many of the early research projects would produce exactly the same results if given the same input; they did not change internally with experience. A chess search, for example, would always lose to the same sequence of moves, and until very recently, this had to be dealt with by the programming team either tweaking some of the search parameters or training the program with a selected database of chess moves that covered the exact situations where the program had failed. Hence the learning was taking place in the experimenters' brains, not inside the AI systems! The idea of the system doing the learning itself was not seen as necessary or feasible.

THE EVOLUTION OF MACHINE LEARNING

Of course, this attitude was not universal; some people did worry about learning. As we saw in chapter 5, Arthur Samuel was very concerned with this for his checker-playing scheme. Samuel realized that you have to learn from your mistakes (for game playing, at least), and actually a lot of learning is done *while* playing the game or carrying out the task.

So what do we mean by learning? Everyday experience suggests that learning is one of the most basic and most vital human characteristics. People remark on the value of learning in all aspects of life: "I've learned how not to antagonize my boss"; "I learned a lot from my project in Germany"; "I can now ride a bike, and solve differential equations," or "Life has taught me how to be successful." The key factor is change; after learning, there has to be a change, either in knowledge, behavior, or beliefs. Notice the breadth of experience involved in these examples: from behavior and relationships, to facts and knowledge, to long-term plans. Most dictionaries capture this broadness by describing learning as something like "a process of acquiring knowledge, skills, attitudes, or values, through study, experience, or teaching, that causes a change of behavior that is persistent and measurable."[1] But whatever is learned, the overarching characteristic of a learning system, unlike conventional computer systems, is that it is able to give different (i.e., better) responses when faced with exactly the same situation as has been experienced previously.

As AI developed, a field called Machine Learning (ML) became established as a branch of it. ML was strictly focused on learning methods and produced a range of very mathematical techniques. For a while, ML almost existed as a separate area from AI, with its own highly respected journal, conferences, and community.[2] Nowadays, ML is everywhere and in everything. Modern learning techniques, using statistical methods and Deep Learning, have driven ML to the fore and opened up new, much wider application areas. ML now selects products for our approval, protects our computers from spam and viruses, detects fraudulent activity in bank and credit card transactions, and monitors data in science, business, and commerce. We increasingly rely on ML for everyday living, just as we all said a few years ago how much we rely on computers. Let's first consider how learning evolved to produce modern computer learning from Big Data, and then look at the amazing revolution in Deep Learning.

DATA MINING IN SUPERMARKETS

Many ML applications are concerned with searching for patterns in data; this has become big business because of the huge amount of data that is now available. The internet has facilitated the collection of customer information to create vast databases. These are also compiled by governments, companies, and other agencies. A good example of ML applied to large databases (often called data mining) is the use of such technology by national and global supermarkets.

For instance, in the UK, Tesco has been top of the list of retailers for a while, with around 10 percent of the UK retail market. This is a big company, considerably larger than the next one on the list. Tesco sells over one-quarter of the UK's groceries and is the biggest private-sector employer, with over 250,000 employees. But our interest in this company is not its size, but its pioneering introduction and successful exploitation of the loyalty card concept (which, incidentally, contributed greatly to its growth). Tesco's loyalty card, known as Clubcard, was introduced in 1995 and is currently used by more than 16 million customers; that's more than the most popular bank card in the UK. Tesco spends well over £20 million per year on the collection, processing, and analysis of data from their loyalty card scheme. This is a central plank in their sales and customer relations strategy and has been remarkably successful. In 2008, analysts reported (Humby, Hunt, and Phillips, 2008): "The Tesco Clubcard is widely regarded as the world's most successful loyalty operation. The program has helped Tesco become the United Kingdom's No. 1 retailer, as well as the world's most successful Internet supermarket, one of

Europe's fastest-growing financial services companies, and arguably one of the globe's biggest exponents of customer relationship management." Of course, loyalty cards are now widely used, and Tesco's fortunes aren't solely based on their card (Tesco posted a £3.8 billion profit in 2011 and a £6.4 billion loss in 2014), but this case study shows how Big Data became so important commercially.

Tesco's stated aim is "to create value for customers to earn their *lifetime* loyalty" italics added.[3] Customers are given vouchers that reward them in proportion to the amount they spend (currently around £300 million is held on members' Clubcards). The secret is in their targeting of the rewards; the vouchers are almost unique—on average, for each of the 16 million customers, any offer will be shared by at most only one other person. This is personalized reward on a massive scale. So how is it done? How can ML be used to do this?

Let's take a small but illustrative example. A small supermarket has built up a large database on its customers. Each customer is uniquely identified through a loyalty card code, and the amount spent, the dates of visits, and the mix of items bought are all recorded when the person uses the card at the checkout. A sample of data is shown in table 7.1. There are many possible uses of customer data, usually all aimed at the enhancement of the business. Let's assume that the supermarket has an in-house café for customers and wants to know if there is any relationship between the customers who use the café and their particular spending activity.

One way of processing the data is to build a decision tree designed to show any pattern in customers' behavior that relates to their use of the

Table 7.1 Data from a small sample of customers.

Customer ID	Spending	Visits	Goods	Café
1	High	Rarely	Food	No
2	Medium	Rarely	Other	Yes
3	Low	Often	Food	No
4	Medium	Often	Food	Yes
5	High	Rarely	Other	No
6	Low	Often	Other	Yes
7	High	Often	Food	Yes
8	Low	Rarely	Food	Yes
9	Medium	Often	Other	Yes
10	Low	Rarely	Food	No

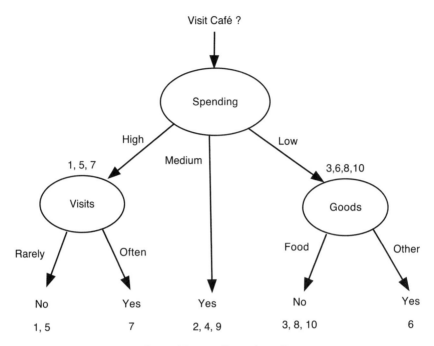

7.1 Deciding if customers from table 7.1 will visit the café.

café. Decision-tree learning scans each data item in turn and generates a decision rule for the desired variable (in this case visiting the café). This is repeated for each outcome of the rule, and thus further rules are generated that refine the decision. In brief, the resulting tree captures any reliable pattern in the data that corresponds with visits to the café. The decision tree learned for this data by an ML program is shown in figure 7.1.

You can see how it works by checking back with the data. The learning algorithm has selected "Spending" as the first decision point in the tree. Looking at the "Spending" column in table 7.1, we see only three cases of Medium spending (customers 2, 4, and 9), and they all go to the café. Hence, with no further ado, the tree decides that if Spending = Medium, then the answer is Yes, they do use the café.

Of the customers that have High spending (1, 5, and 7), the "Visits" data is sufficient to separate the café users from the nonusers. Reading the left branch of the tree, we get: If you are a High spender and you visit the store rarely, then you don't use the café, but if you visit often, then you do use it.

Box 7.1
Café visiting algorithm

Inputs: Spending, Visits, Goods
Output: Yes/No

IF Spending = High THEN, IF Visits = Rarely THEN result = No
                                                                ELSE result = Yes,
IF Spending = Medium THEN result = Yes,
IF Spending = Low THEN, IF Goods = Food THEN result = No
                                                                ELSE result = Yes

---

The remaining decision deals with the Low-spending customers: numbers 3, 6, 8, and 10. Unfortunately, a complete separation of the pattern is impossible with this data, as customers 8 and 10 are in conflict—they are identical except for the café usage. The learning algorithm knows that this cannot be resolved perfectly and makes the decision on the basis of the least wrong number of decisions (8 is in error).

The method does as well as it can; any other tree would have the same number of errors or more, and this is the best result for that particular data. But we can calculate the risks of using a tree by examining any errors. In the Low-spending branch, there are four data cases that were involved, and one produced an error; hence, the error rate is 25 percent and this can be recorded as a risk factor or confidence level for that branch. For the whole decision tree, the error rate is 1 out of 10 (i.e., 10 percent).

LEARNING ALGORITHMS THAT LEARN ALGORITHMS

These generated decision trees can run into hundreds of branches, and then become very unfriendly for humans! But the great thing about this kind of learning is that we don't have to read the trees; a decision tree can be used as a program (i.e., the learning program outputs another program). We can see this by writing the tree in a different way. Box 7.1 contains just the previous tree, restructured using IF THEN rules. You can check the algorithm by following it against figure 7.1.

We can now use this program to make decisions about new data. Table 7.2 shows three new customers who have visited the store, and we can process their data to see if they are café users.

Table 7.2 Some new customers–can we predict if they will use the café?

Customer ID	Spend	Visits	Goods
11	High	Rarely	Other
12	Medium	Rarely	Food
13	Low	Rarely	Food

For customer 11, Spending is High and Visits occur Rarely, so the prediction is that this person won't go to the café. Also, there are no data conflicts, so confidence in this result is high. Customer number 12 is a Medium spender and so the decision is yes, this person is a café user. However, the exact data combination for customer 12 has not been seen in the original data. There are four possible cases that could accompany Medium spending. Three of these cases are in the original data, but this case (Rarely and Food) has not been seen for Medium spending and so could theoretically give a different result. For this reason, the algorithm estimates the "error" as possibly being one out of four (i.e., 25 percent; equivalently, we could say confidence in the result is 75 percent). A different situation occurs with customer 13; the decision is based on Low spending and Food ("Visits" is not used) and gives "No," a noncafé user. But this decision in the tree was built on some conflicting data, as we noticed (lines 8 and 10 in the original data). Consequently, the "No" branch of the decision has one error out of three cases, and the algorithm reports a 33 percent error possibility for using that branch.

This simple example illustrates several features of data mining: ML can autonomously find interesting patterns in large data banks; statistics gleaned from the data can provide estimations of error rates and confidence in the results; and the output of data mining can be algorithms that test other data or look for related patterns. These learning processes can run over the data continuously, updating their conclusions, error rates, or outputs as new data streams come in.

DISCOVERING PATTERNS

These methods can also operate in a discovery mode, looking for interesting but *unspecified* patterns. In this example, we directed the learning toward a particular interest; We wanted to know about café visits, so we selected the Café data as the desired output. But undirected mining is possible, whereby any interesting pattern in the data is detected and reported. Of course, the criteria for "interesting" needs to be specified, but even simple correlations produce surprising and often useful results. Taking the previous customer data from table 7.1 and applying a clustering approach produces a number of possible relationships.

By taking each variable and looking for corresponding patterns or changes in the other variable, we find that the rare visitors spend more on nonfood goods, while the frequent visitors spend more on food items and use the café a lot. Also, the café users tend to be low spenders, and the food buyers are medium spenders. Of course, this is a tiny data set and therefore has no validity; a Big Data source would produce reliable and useful results.

A retailer wants to know what factors relate to increased spending and how they vary across different kinds of customer. For example, high-spending customers might be identified by purchasing luxury products or expensive versions of common items; this group then could be sent special offers on high-end goods; meanwhile, hard-pressed students might receive coupons for bulk buys of basic essentials. Both groups are important because the results of data mining allow the company to optimize its sales strategy for individuals, not just the public at large. This "customer profiling" has been very successful in increasing sales and tempting customers into new ways of spending. The technology serves to "enhance loyalty and increase shopping visits and basket sizes."[4]

It is significant that the software engineers who build these data-mining systems no longer follow traditional programming work patterns. They use a lot of existing software rather than writing new code, and spend a good deal of time on designing the attributes that are to be processed and ensuring that the results are appropriate for the aims of the company.

BIG DATA

Huge stores of data have built up very rapidly in this century—and are growing at astounding rates (prepare for some mind-boggling numbers). For example, in astronomy, ground- and space-based observatories can easily produce one terabyte of data per day—in terms of quantities, that's

Box 7.2
Measuring Big Data

> Consider the typical sizes of some digital images:
> 100 kilobytes (100,000 bytes) = a low-quality web-page image
> 1 megabyte (1,000,000 bytes) = a medium-quality camera image
> 10 megabytes = a high-quality camera image
>
> Let's use the size of a good-quality camera image as a measure, (a still image is an intermediate average between smaller text files and larger video footage). The memory cards in cameras are available in gigabyte sizes, and a gigabyte can store around 100 pictures. (A DVD disc stores 4.7 gigabytes.) Using this analogy, we can translate these terms into more meaningful numbers:
> 10 megabytes = 1 picture
> 1 gigabyte = 100 pictures
> 1 terabyte = 100,000 pictures, or 200 DVDs
> 1 petabyte = 100 million pictures, or 200,000 DVDs
> 1 exabyte = 100 billion pictures, or 200 million DVDs
> 1 zettabyte = 100 trillion pictures or 200 billion DVDs
>
> The terms go up by a multiplier of 1,000: a terabyte is one thousand gigabytes. There are terms for even higher values, which are ready when needed.

about 100,000 good-quality pictures from a digital camera or phone every day. Box 7.2 provides a guide to large quantities.

One of the biggest internet data generators is YouTube, which produces about 100 petabytes per year (10 billion pictures). (The internet itself sends around 50 billion pictures' worth of data every day, although this is traffic and not storage.) The Human Genome Project started a biological data industry that is doubling its data every seven months; this is quite worrying for the biologists, as they now produce over 1 EB (exabytes) a year (i.e., 100 billion pictures, worth of data).

That's just a few examples. Then there are the commercial giants like Google, the information technology (IT) industry, and the rest of the commercial sector, which hoard anything they can on their customers: you and me. Google probably has more data storage capacity than any other single organization. They have over 20 Big Data centers, each costing up to $1 billion, and estimates from power loading suggest they run over 2 million servers, actively storing and dishing out data. This suggests they have at least 15 exabytes of online capacity, not counting magnetic tape archives, and consume around 1,000 megawatts of power.[5] The power consumption of the world's servers already exceed that of countries like the UK. This is a serious issue, and Google has used its own Deep Learning technology in experiments that promise to reduce its server power demand by 40 percent.

Most branches of science seem to be producing vast amounts of data, and storage and processing are becoming major challenges. By 2025, there will be 165 ZB (zettabytes) of raw data available globally. That is, 165 followed by 21 zeroes. These are truly staggering amounts of data that require factory-sized data farms to store and distribute huge data sets, most of which are freely available. It would take more than a lifetime for a person to wade through such volumes, so machines are essential for doing anything with these resources. We will discuss Big Data again in the next chapter.

STATISTICS IS IMPORTANT, BUT MISUNDERSTOOD

Statistical techniques also play a big part in mining information from large data sources. We have just seen how learning algorithms can scan lots of data to find the best decision tree for a task or potentially interesting patterns. We also saw how the statistics of conflicts and other features in data can provide summaries of it as a whole. Statistical techniques and probability theory can come into play when there are imperfections or uncertainties in the data.

Uncertainty means ignorance of one form or another; it means that we don't have the exact knowledge that would enable us to reach a perfectly reliable conclusion. There are various sources of uncertainty, usually involving incomplete, inconsistent, or imprecise information. Examples include missing data (e.g., unknown blood pressure), conflicting data (e.g., pregnancy indicated, but in a person who is male), or vague data (e.g., instrument error is 30 percent)

Probability provides a way of summarizing the uncertainties that come from ignorance. A measure of probability can represent the level of belief in a situation, given the available evidence. Many AI techniques are based on probabilistic methods, but unfortunately, humans are particularly bad at assessing uncertainty and risk. We have various biases that interfere with the cold calculation of risk, and we often judge things incorrectly. And we often jump to quick conclusions based on overly simple reasoning, reliance on faulty assumptions, or ignoring some of the data. This means that the field of statistics is notoriously counterintuitive for humans.

Unlike the abstractions used in much of math and science, nothing is truly independent. So any prior information is likely to be important and should not be ignored. A way of dealing with this extra information was developed by the eighteenth-century clergyman Thomas Bayes. Bayes's theorem describes the probability of an event based on prior knowledge of conditions that might be related to that event.

Without performing detailed calculations, a simplified example demonstrates the value of this approach. Meningitis is a very dangerous infection that often initially causes a stiff neck (in about 50 percent of cases). The chance of getting meningitis is 1 in 50,000, and the chance of getting a stiff neck is about 1 in 20. So, what is the chance of actually having meningitis if you have developed a stiff neck? Applying Bayes's formula, it turns out that only 1 in 5,000 people with a stiff neck can be expected to have meningitis. Bayes's method is easy to implement and forms the basis for AI systems that continuously update beliefs as more evidence becomes available.

THE REVOLUTION CONTINUES—WITH DEEP ZERO

In chapter 6, we saw how Google DeepMind used leep learning to create some very impressive game-playing systems. AlphaGo is significant because it does not use the usual minimax search algorithm, but rather relies on a Monte Carlo tree generator to explore the move possibilities. Monte Carlo tree search (MCTS) is essentially a random generator that estimates the value of branches in the tree and gradually increases the accuracy of the estimates as it explores. It has been proven that MCTS eventually converges on the minimax tree. AlphaGo used deep networks to select branches (policy networks) and to evaluate positions (value networks). These networks were first trained on data from human experts and then refined by self-playing sessions. The success of this system is due to the subtle combining of the networks with the MCTS search process.

In quick succession, Google DeepMind announced AlphaGo Zero. The "Zero" in the name indicates that there is no human input into the learning process: no expert training data, no game knowledge, no supervision—all it knows is the rules of the game, and it starts from scratch. AlphaGo Zero only uses reinforcement learning; it teaches itself by training a neural network by playing games against itself. In a 100-game match against the champion-defeating AlphaGo, it won all the games.

But let's move on to the pinnacle of this series, AlphaZero. By removing all the Go-specific content from AlphaGo Zero, Google DeepMind produced a new system that could, in principle, learn to play any similar board game, given the rules of the game. It has already learned to play several two-person board games: chess, shogi (Japanese chess), and Go. As for AlphaGo Zero, it has no hand-engineered features, no chess knowledge, and uses self-play to learn everything about the game, from basic beginner level on. The performance was again astounding. After only 4 hours of self-learning, AlphaZero played a 100-game match against one of the top chess programs, Stockfish 8, AlphaZero won 28 games and drew 72 games.

Table 7.3 summarizes these DeepMind achievements and, for comparison, includes the Deep Blue chess milestone. AlphaZero has capitalized on the experience of the prior programs and looks like the final stage in this series. A number of significant developments emerge from this research:

Better algorithms AlphaZero uses methods that work and has avoided all human inputs and bottlenecks. It is essentially a novel use of *reinforcement learning*, the very method employed in the Atari DQN system to find connections between the video input, the joystick actions, and the score. This is *end-to-end learning*—that is, only the action (input) and the result (output) are visible to the system; there is no game-specific information, training, or supervision; it has to learn by doing—just as a human player would. Reinforcement learning uses reward signals to learn the correspondence between events and their possible causes even though they may not occur together. In the game of chess the key winning move may be many moves in advance of the final capture of the king.

Table 7.3 The mastery of chess and other games by Deep Learning.

Game	Date	Name	Method	Reference
Chess	1997	Deep Blue	Minimax search. Hand-crafted evaluation. Trained on expert book data.	Hsu (2004)
Atari–49 games	2013	DQN	End-to-end reinforcement learning of action sequences. No features, or training.	Mnih et al. (2015)
Go	2014	CNN	Move prediction by evaluation network. Trained on expert data.	Maddison et al. (2014)
Go	2016	AlphaGo	Monte-Carlo tree search. Policy and value networks Trained on expert & self-play.	Silver et al. (2016)
Go	2017	AlphaGo Zero	Monte-Carlo tree search. No human expert data. No knowledge. Only reinforcement learning.	Silver et al. (2017b)
Chess, shogi, Go	2017	AlphaZero	Monte-Carlo tree search. All learned from scratch through self-play.	Silver et al. (2017a)

Note: A key reference source is given for each system.

By playing games against itself, AlphaZero becomes its own teacher. The system begins knowing nothing about the game. It then plays games against itself, during which a neural network is tuned and updated to predict moves; by this means, the quality of the self-play games increases, leading to more and more accurate neural networks and increasingly stronger versions of AlphaZero.

Less data, no engineering The Big Data requirements for supervised (and unsupervised) learning demand a lot of time and effort. As DeepMind says: expert data is "often expensive, unreliable or simply unavailable" (Silver et al., 2017a). The provision of Big Data files to train a learning system can give an uneasy feeling of an unfair advantage; humans don't have access to such rich and extensive expertise. AlphaZero has completely removed this concern. AlphaGo was trained on a database with 30 million moves, but AlphaZero has no training data. There is no feature engineering, and no supervision of any kind; and the only way it gets any chess knowledge is through experience—by playing the game.

Performance increases The sequence of programs in table 7.3 represent continuing performance leaps. AlphaGo, which was already superior to the excellent move prediction network, the convolutional neural network (CNN), defeated the world champion Go player. AlphaGo Zero then completely demolished AlphaGo in a 100-game match, in which it won all the games. And then AlphaZero defeated AlphaGo Zero at Go, as well as learning and playing chess and shogi against the world-champion programs Stockfish and Elmo, respectively.

Hardware developments The processor chips in most computers are known as Central Processing Units (CPUs), but deep neural networks only perform a great deal of simple arithmetic in large arrays of numbers. It was found that the specialized chips developed for games and other graphic-intensive displays, Graphic Processing Units (GPUs), could do this work more efficiently, and GPUs are often used in deep networks.

Google has created a software tool, known as TensorFlow, especially for handling multiple arrays of numerical data, and they now have custom made chips, inside special Tensor Processing Units (TPUs) tailored to run TensorFlow at high speeds.

AlphaZero used 4 TPUs when playing matches, but over 5,000 TPUs were involved in the learning process. These TPUs are running in Google's data centers and they are being made available as a cloud-based service.

The reactions to AlphaZero's performance were positive and reflective. Demis Hassabis, of Google DeepMind, commenting on removing the need to learn from humans, said that we are "no longer constrained by the limits of human knowledge" (Silver et al., 2017a).[6] Others were impressed that "it figured everything out from scratch" and noted that this is both "scary and promising." It is promising because this removes another bottleneck—the handcrafting involved in training—and scary because we don't know the limits of this learning method.

Apparently, AlphaZero's playing style is different from other computer systems. It is more focused; it tends to discard unpromising avenues rapidly and concentrates on deeper analysis of interesting positions, rather like human behavior. This is seen in its examination of only 80,000 positions per second, while the more conventional Stockfish looks at 70 million per second. Also, it plays better chess with the white pieces (but for Go and shogi, it played as well with either side). This is surprising; the white pieces are associated with a slight advantage in chess, but not enough to explain the results. Maybe the real reason is that AlphaZero just needs more self-playing time; it possibly tends to explore the "winning" configurations first. Google DeepMind says their plan now is to explore how this learning technology can be exploited in useful application areas.

OBSERVATIONS

- Learning has moved center stage for AI, after long neglect as a sideline.
- Machine Learning (ML) and Big Data created a new industry called *data mining*. However, terms like *Big Data* and *Data Mining* are going out of fashion—it's all called *AI* now.
- We no longer have to write software to locate items in data stores. We can just specify patterns of interest, and learning algorithms will then produce software that finds relevant items.
- Uncertainty represents ignorance, which is unavoidable in knowledge-based systems. Statistical and probabilistic methods provide ways of handling uncertainty and combining evidence.
- Probabilities can summarize beliefs, but beware of counterintuitive results.
- Bayesian methods are very effective and are used in most modern AI systems that reason with probabilistic information. A crucial feature of the Bayes theorem is that it can continuously take in evidence and revise its estimates.
- Risk-assessment methods are very important in real applications. All engineering and most big science projects spend a great deal of time and money on risk analysis.
- The Google DeepMind research on deep game playing has achieved a new breakthrough with its AlphaZero system. AlphaZero is a general game-learner that learns to play chess or Go with no external training or human or data input. This represents a new form of learning from scratch, essentially end-to-end, trial-and-error learning. Being free of human training, it is "no longer constrained by the limits of human knowledge."

8

DEEP THOUGHT AND OTHER ORACLES

Integrity without knowledge is weak and useless, and knowledge without integrity is dangerous and dreadful.
—Samuel Johnson, *The History of Rasselas, Prince of Abissinia*, 1759

So what *is* wrong with artificial intelligence (AI)? Given the recent amazing advances and continuing healthy progress in AI, as outlined in the previous chapters, why does Part I of this book have such a critical title? First, let's make it clear that there is nothing wrong with AI for many purposes. It is producing powerful and valuable applications in all areas where computers are used—and finding many new applications as well. In recent years, we have seen notable growth in activity and maturity and some truly impressive results and applications. Thus, we have advances in areas like real-time face recognition, medical image interpretation, autonomous vehicle control, speech-based interaction and communication, planning, and diagnosis systems. AI is being used everywhere, from archaeology to supply-chain management.

The problems come when social aspects become important and humanlike performance is required. Earlier in this book, I suggested that intelligence exists in three forms: the mental abilities of individuals; the social interaction of several participants; and the collective knowledge generated and recorded through human cultural activity. The first and last of these can readily be found in AI. Modern AI applications often solve particular problems and thus appear to operate like an individual agent, while cultural intelligence is available through the processing of the enormous resources of Big Data. But the social side—using AI to interact with humans—seems to lag. It's not that social interactions are not being studied (they are under active research), but I contend that AI is insufficient on its own to provide human-level performance in the social arena. In other words, individual AI agents are fundamentally inadequate for long-term social interactions with humans. This chapter explains why that is.

AI IS A HIGHLY FOCUSED BUSINESS

AI is a high-tech business, and as soon as a research laboratory produces a working prototype, there is a rush to build applications and deliver commercial products to users. Most of the early research was not stable enough to go to market, but the field has now advanced so much that reliable and professional performance is normal for applications in well-defined areas. In fact, AI has often been a victim of its own success and doesn't always get the recognition it deserves. It usually gets incorporated into modern software or high-tech systems at some stage, and as soon as this happens, it tends not to be known as AI anymore. For example, your computer's email spam filter and virus checker are both using AI techniques to learn about message patterns and deal with items that you consider to be junk. And your digital camera is using AI image-processing methods to locate targets, reduce noise, and monitor a range of camera functions to improve the pictures. Clever reasoning and inference techniques are being used to analyze and detect patterns in all kinds of large databases to help the police by finding connections in crime data, to monitor credit card usage patterns to detect fraudulent behavior, and to produce tools for biologists to help their work in genetics, medicine, and the understanding of cellular life itself.

One of the major achievements of computer science is the development of specifications and tools for defining and resolving problems. This discipline, known as *software engineering*, covers the professional development of software to produce high-quality products, very similar to the processes used by other engineering professionals to build railways, cars, planes, phones, computers, and practically everything else that is manufactured. The use of engineering techniques, including good project management, allows the uncertainties and unknowns in an application area to be identified and managed. By specifying the task to be achieved, the range of operations and the desired behavior can be specified in a performance goal. We can characterize this approach as being *task-based*, in that the programmer has aimed (implicitly or explicitly) at a specific goal or target performance. This methodology gives a degree of focus and reduces some of the application uncertainties, thus helping the designer to optimize performance and also assisting in commercialization. Crucially, this also explains why AI programs are very skilled in their particular application task (which we might say requires intelligence), but they are completely unable to transfer their skill outside their task domain. In chapter 2, we noted the same skill transfer issue for robots.

TASK-BASED AI

Given this insight, we can appreciate why AI systems can be very powerful in a specific area, but completely inappropriate for anything beyond that scope. This kind of AI has been called *weak AI*, but we will use the term *task-based AI* instead.

In contrast, *strong AI* refers to the concept of *general intelligence*; that is, a non-task-specific but general capability that can be applied in whatever circumstances arise. Of course, general intelligence is what humans display, and *Artificial General Intelligence (AGI)* is the ultimate goal of AI research. (Google DeepMind says its mission is to "Solve Intelligence"— but more on this later.) However, there have been very few attempts to build general intelligence machines, and they have always been disappointing. Indeed, the many and varied successes of AI can be attributed to the fact that they are task-based, and thus have clear achievement goals. Objections may be raised that people also become specialized in their mental abilities; for example, chess grandmasters aren't necessarily well-rounded, flexible individuals who are as clever in other areas—some have had rather dysfunctional lives.[1] But the point is not so much that a machine cannot do more tasks, it is that we, the designers, do not know how to make them more flexible and general.

MACHINE ORACLES

Machines with general intelligence have stimulated the imaginations of many science fiction authors. An entertaining example is the computer Deep Thought, in Douglas Adams's comic novel *The Hitch-hiker's Guide to the Galaxy*.[2] Deep Thought (no connection to the chess program) is a computer oracle (the size of a small city) that is designed to answer any question. In the story, the main characters ask it for an answer to "The Ultimate Question of Life, the Universe, and Everything," and it takes seven-and-a-half million years to produce the answer: 42. (Deep Thought then says that to make sense of the answer, the forgotten question needs to be known, so it designs yet another computer to calculate the question.)

A much more realistic example is the computer HAL 9000 in Stanley Kubrick's 1968 film *2001: A Space Odyssey*.[3] HAL was the main control computer for a deep space vehicle and was omniscient in that "he"[4] had command over all the ship's systems, including the crew's environment and safety. HAL also had conversations with the crew and played chess with them. This film was hugely influential because it was so realistic and professional in the depiction of the space technology. For such an early production—before the internet, personal computers, Global Positioning System (GPS), or mobile phones—even the computer displays were impressive mock-ups of what was to come. Without modern computer graphics technology like computer-generated imagery (CGI), the many computer displays were built on the movie sets using back projection and other optical tricks.

The plot, involving HAL, raised interesting ethical questions about the ability of advanced, intelligent computers to reason on their own, and even counter the commands of their human operators. HAL believed he was "foolproof and incapable of error," and this belief led to a reasoning sequence that was fatal for most of the crew. The last remaining astronaut disabled HAL's circuits in self-defense.

Many of the AI ideas featured in HAL have come to pass in various forms in modern software. The excellent book *HAL's Legacy* (Stork, 1998) consists of chapters by relevant experts who assess the computer's various AI abilities, which include speech production, speech recognition, face recognition (including lip reading, in a crucial scene in the movie), creativity, understanding emotional behaviors, automated reasoning (including diagnosis and prediction), and, of course, playing chess.[5]

These fictional examples of general intelligence perform rather like oracles, in that they don't have robot bodies, but they interact with humans through fixed computer interfaces (usually speech or text) and mainly operate in a question-and-answer mode. As mentioned in an earlier chapter, Alan Turing also considered the problem of intelligence. He argued against definitions and instead proposed an "imitation game," wherein a computer had to imitate a human during a question-and-answer session. If the human questioner could not tell that his or her conversational partner was a computer within a certain time, then it had passed the test. The details of this game, now known as the Turing test, together with a range of objections and other issues, are still available in Turing's very readable paper, *Computing Machinery and Intelligence* (Turing, 1950).[6]

Obviously, the computer must not be seen by the interrogator, so some form of intermediary is necessary for the interactions. This is usually achieved by passing textual questions and answers via a computer display and keyboard. However, Turing made it very clear that the physical aspects were of no interest (1950, 434):

The new problem has the advantage of drawing a fairly sharp line between the physical and the intellectual capacities of a man. No engineer or chemist claims to be able to produce a material which is indistinguishable from the human skin. It is possible that at some time this might be done, but even supposing this invention available we should feel there was little point in trying to make a "thinking machine" more human by dressing it up in such artificial flesh. *The form in which we have set the problem reflects this fact* in the condition which prevents the interrogator from seeing or touching the other competitors, or hearing their voices. (italics mine).

He recognized that this would restrict matters to higher-level thoughts (Turing, 1950, 458, 456):

> The imperatives that can be obeyed by a machine that has no limbs are bound to be of a rather intellectual character…

and

> It will not, for instance, be provided with legs, so that it could not be asked to go out and fill the coal scuttle. Possibly it might not have eyes.

This is important because Turing saw the imitation game as a test of pure intellectual skills and knowledge, and thereby deprecated highly subjective questions concerning personal, body-centered, sensorimotor experience. Turing's interrogator did ask, "What is the length of your hair?" but this is a factual kind of question that might be answered by looking up a database to see what is a typical length of hair for a human. It would be much harder to answer questions about a toothache, for instance. The machine might report that its last toothache felt "awful" or "very painful" by looking at reported records, but a little more probing would expose the lack of subjectivity, the lack of *ownership* of the experience that cannot be simulated from purely factual data. Probably a clever program would bring in evasive remarks at this point, like "I can't describe how bad it was."

Another key point that has not always been recognized is that in the original game, the interrogator had to interact with both a human and a machine and decide which was which. In the modern version, there is only one subject, and the interrogator has to decide whether it is a human or a machine. The original game had the computer and human together in the same hidden room, so they could both see all the exchanges, and the human was allowed to help the interrogator. This setup gives a much richer interaction among the three participants, as the computer and human can comment to the interrogator about the other's behavior. So the human player might say, "That's wrong, no human would say that. I'm the human!" This level of exchange and interaction would make the test even harder than the current question-and-answer sessions, with their two-way format.

The Turing test has been discussed extensively, and it has stimulated the implementation of many programs that try to deceive their interrogators. This has been enshrined in the Loebner Prize contest, an annual tournament started in 1991, with significant cash prizes. In June 2014, on the 60th anniversary of Turing's death, a program with the backstory of a simulated Russian teenager was declared to have passed the Turing test at the Royal Society in London.[7] It was claimed that the program had passed the test because a third of the judges thought that they had been communicating with a human, and Turing said he'd be happy when "an average interrogator will not have more than 70% chance of making the right identification after five minutes of questioning" (1950, 442). However, a great deal of confusion and controversy surrounds this issue. Many can't agree on the criteria, and others see the test as much more than a game. Professor Stevan Harnad, a cognitive scientist at the Universities of Southampton and Quebec in Montreal, said that whatever had happened at the Royal Society, it did not amount to passing the Turing test. 'It's nonsense, complete nonsense," he said. 'We have not passed the Turing test. We are not even close."[8] I have to agree; the level of conversation in these tests was not high.

This long-running saga shows that simulating human behavior is extremely difficult. Although Turing makes it quite clear that the "game" only applies to "purely intellectual fields" (Turing and Copeland, 2004, 463), even this restriction admits a wide range of abstract ideas, puzzles, thoughts about facts, deductions, inferences, and other elements. And for each one of these, an exchange could be about the wisdom of a certain choice and the possible alternatives. Writing a program that could do this in the way that a human would talk about it is much harder than just producing the solution to a formal problem. At any rate, along with the rise of avatars, interacting applications have become quite popular, and the internet offers a range of "chatbots" that you can try out and see what these question-and-answer sessions are actually like.[9] Just don't expect too much!

It is strange that people are so eager to see this test passed. Just as with chess, we can ask: What does it mean to pass this milestone? When a program can maintain a conversation long enough to fool someone into thinking it is a human, what does that mean? The milestone is even more arbitrary than for chess, which does have a very precise criterion. There is no doubt that more advanced programs will be written, and performance will continually improve. But increasing the number of people fooled does not amount to a breakthrough. It is clear from his writing that

Turing would be happy to say that a chess computer was "thinking" about chess. He also had much to say about how a program could learn the things necessary to simulate a human. So he was clearly not too bothered about thresholds and criteria. It is significant that Turing's 1950 paper was written in a very different style from the formal, mathematical approach of his other work; it is amusing and ironic in places, almost sardonic in some of his refutations of the nine arguments that he proposes as opposing views (Turing and Copeland, 2004, 433) So all this suggests that we should not take such tests too seriously, and just enjoy the ingenuity and entertainment that creative programming offers.

KNOWLEDGE ENGINEERING

As an example of an oracle, we can look at question-answering systems. There are many situations where exchanges follow a question-and-answer pattern. A help-line service run from a call center is a commercial example, and general-knowledge quizzes are very popular recreational examples. The difference between these question-and-answer systems and general Turing test candidates is that the interaction is structured in a more formal question-and-answer format, and usually the topic area for the questions is limited to a particular domain. This topic scope is called the *domain of discourse*. Thus, a call center might advise customers on train services; acceptable questions would involve destinations, schedules, and costs; and any questions outside the domain of discourse (like "What is the weather like in Manchester?") do not have to be answered (but the response is a polite apology for ignorance).

In 2007, IBM started a project called DeepQA, with the aim of building a powerful question-answering system. The domain of discourse was general knowledge, so this was a very challenging project. The result was Watson, a computer system specifically developed to answer questions, posed in natural language, on the quiz show *Jeopardy!* IBM collaborated with several universities on the AI technology involved, and in 2011, Watson competed on *Jeopardy!* against two former winners and won first place, with a prize of $1 million. Watson worked by analyzing the questions and searching over 200 million pages of data, including the full text

Table 8.1 A few interactive tasks in terms of environmental complexity, with chess as the comparison.

Application	Level of chaos				Score
	Uncertainty	Novel events	Available knowledge	Order in the task	
Chess	Negligible	Negligible	High	Maximum	6
Help service	Medium	Medium	High	High	14
Watson	High	Medium	High	High	15
Chatbot	High	High	Medium	Medium	18
Turing test	Very high	Very high	Medium	Low	21

of Wikipedia, to find relevant facts and formulate for the answers. It was not connected to the internet during the game, but a complete download of Wikipedia content is a pretty useful resource.

This was a significant research achievement, particularly in the natural-language processing of the questions and the searching of the internal database to formulate a response. Such technology has proved more commercial than chess playing, and IBM have now found applications for the Watson technology in health care, finance, telecommunications, and biotechnology. The ability to communicate in natural language about the topics covered in an application knowledge base, as well as the relatively small size and increasing power of Watson, create many commercial opportunities. Watson is an interesting and developing system, and we will look at it again in chapter 16.

Table 8.1 shows how the earlier idea of analyzing chaos in the environment can be used for both AI and robotics. For the oracle systems described here, their operating environment is essentially the content of the interactive input-output messages. So we can view chaos in this environment in terms of the lack of reliability and disturbances in the message content, plus the degree of available knowledge that is directly relevant to the process, plus the degree of structure in the interactions. All these factors impinge on the difficulties surrounding the task and successful implementation.

From a few different examples, we can see how the difficulties grow as the complexity involved in the interactions increases. Chess is shown in the table just as a comparison with a noninteractive AI application. It doesn't quite manage the minimum score, as players don't have access to complete knowledge of the game; there are always parts of the search tree that they haven't have the time to explore.[10]

Computer-based help services can be as simple as the designer wishes– but not too simple! There is a trade-off between limiting the exchanges to very simple responses and achieving customer satisfaction. We have all experienced phone services that are completely frustrating because the system won't recognize or understand what we are trying to say. Nevertheless, with a large knowledge base and modern natural-language software, very impressive automated services are now possible and widely used. The key to a successful help application is a well-focused topic of discussion and very good replication of human conversational patterns.

The Watson system faces a more challenging task as the range of questions is opened up to much wider and more complex topics. It is, however, based on factual and reasoned topics about general knowledge.

Completely unconstrained interactions offer the greatest challenge. Chatbots are intended to be amusing entertainment, somewhat less serious than the Turing test, which is more difficult because the questions are deliberately designed to expose weaknesses. The inclusion of personal questions in the dialogues for these systems raises the difficulty level.

The nature of the material and the scope of the interactions in a conversational system can be reflected in the domain of discourse. This defines the limits of the "universe" within which the system works and is another way of expressing the complexity or difficulty involved. For the examples from table 8.1, table 8.2 lists their domains of discourse, and this reflects and illustrates the increasing difficulty as the domain expands.

Table 8.2 The constraints on domains of discourse for various tasks.

Application	Domain of discourse
Chess	None
Help service	Constrained to defined application tasks
Watson	Quiz format—facts and general knowledge
Chatbot	Unconstrained—entertainment
Turing test	Unconstrained—competitive

SOCIAL CONVERSATION

Most models of social interaction used in AI assume a kind of alternating statement-and-response structure, but real human conversations are much more complex than that. AI systems rely on access to huge data resources for much of their power, and questions can be rapidly answered by referring to extensive data on the topic of concern. Systems like Apple's Siri and the Google Assistant do this via the cloud using complex networks of knowledge. They are able to look through millions of recorded human conversations and construct a response from similar situations. But note that this "knowledge" is a kind of factual data—it is static, compiled, and recorded. Such systems are fine for understanding commands and looking up facts (they can consult sources such as Wikipedia), and some basic protocol rules will provide acceptable interactions for simple spoken assistants. But they don't relate to the individual experience of the participants.

Many exchanges in meaningful conversations concern our subjective experiences; what we think and feel about things. And feelings can have many dimensions. Music, for example, is commonly experienced through listening, but reading music and playing instruments are other modes of experience, as are composing, or even constructing musical instruments. All these can be discussed in conversation, but for a fruitful interaction, it is necessary for the participants to be receptive to each other's personal experience. It is one thing to know how a piece of music is structured; it is quite another to have experienced the structuring process (only composers can discuss this).

Social conversations are dynamic, unique, and heavily contextualized. You cannot predict the next exchange in an open conversation, and you cannot generate personal experience from data files, which are essentially compilations of past experiences. This is because each human has a unique personality, based on her or his own unique past history, and consequently, each conversation is also unique.

The issue is really is about the limitations of language. Humans have a rich mental life and communicate their thoughts to others through the medium of language. Words are a kind of code: They compress the thoughts, ideas, intentions, and feelings of the speaker into a message that listeners receive and then expand back into thoughts and ideas in their own mental universe. The code (language) is much more restricted than the internal thoughts and experiences of the participants.

Although language has reached sophisticated levels of expression, it can be seen as a serial, unidimensional channel compared with the parallel, multidimensional activity of the human mind. To keep the coding to a manageable size, we rely a great deal on shared experience. The common properties of life are assumed to apply to all of us, and this reduces the necessary content of conversation. Expert wordsmiths, such as the great writers, are able to create rich mental imagery that most of us appreciate and wish we could emulate. The individual, personal concerns in these scenarios are a major difficulty for AI systems, which don't have anything to match the shared experience of the mind.

A good illustration is the problem of pronouns. In the statement: "Jane gave Joan some candy because she was hungry," we can ask "Who was hungry?" and most people will answer "Joan." But if we change the *was* to *wasn't* and ask "Who wasn't hungry?" the answer will be "Jane." In both cases, we are asking who does the pronoun *she* refer to, and the answer changes with subtle changes to the question. There are many of these cases, known as *Winograd schemas*, and they are usually beyond the ability of AI systems to resolve properly (Levesque, Davis, and Morgenstern, 2012).

Here's another example: "The table won't fit through the doorway because it is too wide (or narrow)." When we ask what is too wide or narrow, the answer changes between the table or the doorway.[11] AI systems attempt to deal with such cases by looking up text corpuses and finding previous usage, but pronouns often reference implicit items that are understood by the participants in the context of the mental scenarios they jointly create.[12]

The second example shows how we use our common sense and background knowledge to pick up linguistic meaning. We know that (usually!) pegs don't go in holes when the peg is bigger than the hole.[13] There is no way for an AI system to know this without (1) being given the statement as a fact in some form, or (2) actually experiencing this in the real world (as humans do).

So AI social systems *simulate* experience rather than engage in the real thing. Any extended questioning will soon reveal cracks in their fake subjectivity. Sure, they can be programmed with basic models of dialogue; we all need protocols for efficient and courteous social interactions. However, using past data, even generalized or averaged in some way by learning processes, will not satisfy the needs of a human social partner. Social interactions occur among personalities: agents who have preferences and independent viewpoints based on subjective experiences of themselves as "selves." AI systems do not have this sense of self; they experience all events as factual data.

If we are to make any progress in this area, there are two possibilities: Either our AI systems must somehow be given an enormous amount of background knowledge, or they must be embodied and engage with the world. The former approach involves the encoding of common-sense knowledge as described in the Cyc project in chapter 5. The robotics approach is more recent, and in my view very promising. Social robots will need real-time, real-life experience, so they have to have a body. The body is not just an input-output device; it is the source of all experience that shapes cognitive processes. You have to possess a body, know how to use it, and be prepared to talk about it.

It's extraordinary that people seem happy to ignore the fundamental differences between machines and living systems. Machines are not organisms; they are not alive and thus can never have a meaningful concept of their own death. Parts will wear out in time, and machines can be scrapped and disassembled, or, like Trigger's broom, their parts can be replaced indefinitely.[14] Consequently, although they may come to understand the basics of human life and death, these facts are meaningless for their own situation. Interestingly, if life and death do not exist for robots, then it will be unnecessary to implement analogs of pain and fear. Therefore, they will be less emotional and more platonic than humans, and we will perceive them as useful and skilled artificial assistants.

We live in a human world, not a digital world, and robots cannot fully understand our world. Furthermore, we must organize our digital technology according to our best interests, not the other way around.

OBSERVATIONS

- Intelligence is a product of the human brain, but not an individual human brain in isolation. Social activities are an essential part of the development and functioning of the individual and contribute toward intelligence.
- Recorded data, in all forms, is both a product of and an input into human intelligence.
- AI is ubiquitous in modern software. It is often embedded into systems and has become part of the toolkit of modern software engineering.
- All successful AI systems are designed for a task. Task-based or application-based AI is the norm. Non-task-specific general intelligence, known as artificial general intelligence (AGI) is the ultimate AI goal, but it remains a research dream.
- Systems that converse with humans, answer questions, or try to pass the Turing test can be rated by the degree of constraint on their domain of discourse. Those with a constrained "world" are much easier to design and implement than those with unlimited scope across the universal range of human knowledge and expertise.
- Explicit knowledge in the form of text or documents can now be processed very rapidly by AI to extract patterns, locate meaningful information, and translate into other languages. Huge digital corpuses are now available that cover much of human textual and spoken language.
- General conversations with humans are difficult for AI because they require an understanding of the agent as a "self." All individuals have different personal experiences that have to be learned and cannot be generalized. Self-models are necessary for empathetic understanding.
- Social interactions cannot be simulated by constructing responses from stored data. Real-life interactions are based on personal experience and are deeply grounded and uniquely contextualized in the subjective life-history of the individual agent.
- Language is much more than the tokens that are spoken or recorded in text. Communication requires agents with shared knowledge gained through common experience and learning.

9

BUILDING GIANT BRAINS

"More power, Igor, more power!"
—Often wrongly attributed to Mary Wollstonecraft Shelley.
Dr. Frankenstein had no assistant in Shelley's 1818 novel.

If you want to build an artificially intelligent machine, why not build an artificial brain? This is an obvious plan: The human brain is the source of human intelligence, and so if we could copy it somehow, we would expect to reproduce at least some aspects of intelligence. This attractive approach has been tried before, and there are several ongoing projects in this area. The big snag with "copying the brain somehow" is that no one has yet figured out how the brain actually works. Modern neuroscience has made rapid progress in recent years, but despite gaining much understanding of specific detail, the big picture remains elusive. We still can't answer questions such as, "How is thought generated and held in the brain?" or "How did I store that memory or make that decision?"

Nevertheless, brain science provides a huge source of inspiration for robotics and artificial intelligence (AI). Let's start with a look at brain modeling, and then return to AI to see how artificial neural networks (ANNs) actually work.

It has long been known that the brain consists of a large number of cells, known as *neurons*, which are heavily interconnected and function electrically. Neurons take different shapes and forms, but all seem to work on the basic principle of sending electrical pulses to other neurons across a very complicated network. The number of neurons and interconnections in the brain is very large, at up to 100 billion for the whole nervous system, with around 20 billion in the cerebral cortex. The complexity really shows when you consider that the number of connections for *each* neuron averages at around 1,000 other neurons. These connections between neurons, known as *synapses*, are complex chemical junctions that allow the strength of the connection to vary.

This electrical network is similar to computer hardware, at least to the extent that both have large numbers of interconnected computing elements (neurons in one, transistors in the other) and transmit signals as electrical pulses. This rather gross level of similarity has nevertheless been seen as a compelling analogy by generations of scientists and stimulated a whole subfield of AI that builds brainlike networks.

BRAIN-BUILDING PROJECTS

One of the most ambitious neural network projects of recent years is the Human Brain Project (HBP), headquarters based in EPFL in Switzerland. This 10-year project, which started in October 2013, is building a research infrastructure to help advance neuroscience, medicine, and computing. The HBP employs about 500 scientists at more than 100 institutions across Europe. HBP is funded by the European Commission under the Future Emerging Technologies (FET) program. Only one other project has been funded at a similar level (1 billion euros); this was: the project to develop graphene, the wonder material.[1]

The origins of the HBP was a computer model of a cortical column micro-circuit that ran on an IBM BlueGene supercomputer at the École Polytechnique Fédérale de Lausanne (EPFL) in Switzerland. The outputs of the model were compared with stimulated segments of the rat cortex. A similar project at IBM reported on a mouse model consisting of 8 million neurons and 50 billion synapses and running at one-seventh real-time speed.

The goal of the HBP is to advance our understanding of the human brain. This includes new tools for neuroscience and medicine and new emerging technologies "with brain-like intelligence."[2] A key component of the project is the computing resources, which need to be superpowerful–at least 1,000 times faster than those used in 2013.

The project got off to a bad start, with various criticisms and an open letter signed by 800 scientists. The main complaint was that the aim was so ambitious as to be unscientific, and it was also badly organized and controlled. A mediation team was brought in and the project was reorganized following various reviews in 2015.[3] Although the HBP was criticized by the mediators for raising unrealistic expectations with regard to understanding the brain, even the critics supported the project's parallel goals of delivering computational tools, data integration, and mathematical models for neurological research.

The United States responded to the European effort by starting the Brain Initiative. This has a similar level of funding but is drawn from existing funds across a range of agencies and large corporations. It is less centralized and is more collaborative and distributed, involving many existing US brain research laboratories.

These huge brain models are complex systems that, when they are running simulations of parts of brains, will generate vast amounts of data that must be stored somewhere. The HBP is projected to process data in the terabyte to petabyte range. Their data center specifications indicate that up to 50 petabytes of storage are required.

A good example of how to deal with Big Data is CERN's Large Hadron Collider (LHC), the world's largest particle physics experimental machine. CERN (the European Organization for Nuclear Research) realized early on that the LHC would produce vast amounts of data from its subatomic collision experiments.

They figured that a year of collisions from a single LHC experiment would generate close to 1,000 exabytes of raw data (100,000 billion pictures). To put this in perspective, if you use a typical backup disc for your laptop pictures (say 1 terabyte), you would need to order 1 billion of them to store this experimental data. Moreover, at about 10 watts per drive, you would need 10,000 megawatts of electrical power for them; that's about 1 percent of the world's generation capacity.

CERN has an excellent track record of computer science innovation (the World Wide Web was created there by Sir Tim Berners-Lee), and it planned an integrated grid of global resources to handle the data. The CERN LHC grid consists of around 150 petabytes of storage, distributed across more than 150 computing centers in 36 countries. This is ample because it also built its own sampling technology to compress the data, very impressively, to only 25 petabytes of data each year (2.5 billion pictures). These are the sort of storage tricks that the HBP and other similar large projects have to adopt.

WHOLE BRAIN EMULATION (WBE)

The idea of simulating a complete brain on a supercomputer system has long been an AI dream. We don't yet know how the brain works—but does that matter? We can scan human brains to get a pretty good map of the wiring diagram, and we know quite a bit about individual neurons, so if we simulate all this in a large-enough computer, won't we get a working artificial brain? Various projects have worked on variations of this idea (see de Garis, Shuo, Goertzel, and Ruiting, 2010, for a review of seven studies, including the precursor to HBP, the Blue Brain Project). But it has never been realized in its full-blown form, known as Whole Brain Emulation (WBE). Simulations of small parts of the brain are regularly performed, but WBE is another matter. Unfortunately, many projects are a bit unclear on their aims in this respect; the HBP says, "Can you imagine a brain and its workings being replicated on a computer? That is what the Brain Simulation Platform (BSP) aims to do."[4]

In theory, WBE is possible. There is a principle known as the *universality of computation* that, according to the physicist David Deutsch, "entails that everything that the laws of physics require a physical object to do can, in principle, be emulated in arbitrarily fine detail by some program on a general-purpose computer, provided it is given enough time and memory."[5] This means that the brain, being a finite physical system, could definitely be simulated by a computer—in principle, at least. The neurons in the brain could be simulated, right down to the molecular level if necessary, and a complete working copy could be created. However, the degree of computing power required to emulate the real-time response rate of the human brain is completely beyond our global capacities, current or future.

In addition to the intractable scale of the problem, we have other nontrivial difficulties: namely, our lack of knowledge about the brain's workings and our inability to comprehend and manage such a task. Many scientists consider WBE to be a diversion that will not give a good scientific return on the effort and resources needed. Let's consider just three issues around a large-scale, simulated brain.

The first problem could be called the *interpretation problem*. The brain is a massively parallel system, in which billions of neurons are firing at any one time. So any thought pattern will involve the *collective* pattern of activity of several million neurons. In real brains, we can record the activity of only a few neurons at a time, but we can get access to millions of neurons *simultaneously* in a simulated brain model.

How can we think about or make sense of a pattern that involves millions of changing variables? This is a problem of understanding: The parallel complexity of the brain seems beyond the interpretational abilities of another brain. As Jack Cohen and Ian Stewart quipped: "If our brains were simple enough for us to understand them, we'd be so simple that we couldn't" (Cohen and Stewart, 1995, 8). Their entertaining book contains many interesting insights into complexity and the brain.

Some have suggested that a new mathematics might be developed, which could handle the special difficulties of relating and interpreting millions of simultaneous events. But despite years of necessity, the mother of invention has produced little, and there are few signs of such mathematical breakthroughs on the horizon. Advanced tools are being developed, and they will help somewhat by displaying visualizations of massive neural activity in a similar way to that of medical scanners. But this interpretation issue is more acute with the brain than with other projects with massive data sources, like genome typing, astronomical sensors, or particle physics experiments, because at least with those subjects, we have some reasonable underlying theory that gives a framework for grounding our interpretation and understanding.[6]

A second technical problem is how to ensure that a simulated brain only operates in ways that correspond to real brain states in humans. When a whole-brain model is first switched on, it has no experience, and therefore no memories, so what is it doing? Is it feeling sensations? Is it thinking? What is it thinking? We can call this the *switching-on problem*.

It is known that complex dynamic systems can be very sensitive to the initial values given to their various parameters. A system started with one set of values will perform totally differently if started with another set.

A study of the brain at a clinical level has to correlate the internal measurements gleaned by scanners and neural probes with the activity of the person at the time. Thus, speech, vision, and memory might be the behavioral correlates that are being examined in a live brain experiment. In a simulated system, given that we are aiming for high authenticity, we require the brain model inputs and outputs to be similarly authentic. This is the third problem. For example, it won't be enough to use a camera as a vision system because the properties of a digital camera don't relate to the complex musculature, optical system, and sensory and neural structure of the human eyeball. This is true for other sensory and motor systems. We get signals from tactile skin contact, proprioceptive messages from our skeletal coordinates, temperature and pain, vision and hearing, and other inputs—all highly sophisticated mechanisms in their own right, and all feeding preprocessed signals into the brain. Therefore, the forms of the inputs and outputs, to and from the simulated brain, require as much care and attention in simulation as the brain itself.

THE BRAIN IS A MACHINE—SO WHAT?

Being machines does not belittle us, unless you have a deeply improverished sense of what it can mean to be a machine.
—Grand (2004, 238).

The media and social culture often promote strong positions without much in the way of technical or sound evidence. Social media sites have such a broad influence that they can propagate false propositions, and even bias funding and policymaking. It is worth expanding on the previous section to see why brain-building ideas and proposals need a thorough examination through the pragmatic lens of engineering and scientific evidence.[7]

Gary Marcus, author and professor of psychology and neural science at New York University, wrote an article in *The New York Times* entitled "Face It, Your Brain Is a Computer."[8] In this short article, Marcus tries to reinforce the computational theory of mind, which views the brain as a computer and the mind as the effect of a computation running in the brain.

This is just one example of the many voices shouting out the mantra: *The brain is a machine*. This claim is actually a vacuous truth, a *sine dicendo*. It is a statement that adds nothing and may even be damaging. Modern science is based on reductionist, mechanistic models, which prove extremely effective for understanding most physics and chemistry, and much biology. So it's not difficult to be persuaded that biological systems might be machines of some kind. If we accept that viruses, bacteria, and living cells are miniature machines, and we certainly find mechanistic explanations of their function and behavior efficacious in biology and medicine, then whole systems built from these cells are machines too. Brains, animals, and plants are all machines from this perspective.

However, this view does not carry any useful information. It's like saying that the sea is made of oxygen and hydrogen, with a bit of sodium chloride. What about the behavior of the waves, the different states and conditions of the sea, the forces involved, and the way that large volumes of those molecules interact with all the other entities they influence: the land, atmosphere, and planetary motion? Such empty statements carry no explanatory power, no real information, so they are of no real use. Knowing that a tulip is a cellular machine gives no insight, no guidance for experiments, and certainly no help in building one.

So the machine metaphor offers next to nothing in practical terms. But unfortunately, it is not entirely neutral in effect. It subtly suggests that it is (all too) easy to build a similar machine to the brain, and it also hints that the brain is much easier for science to understand than it actually is. It also ignores the crucial role of context, without which humans (and other biological machines) cannot exist. Let's take each of these points in turn.

The machine metaphor seems to say, "We know how to analyze and build all kinds of machines, and therefore we can build a machine that operates and behaves identically to the brain." However, this assumes that scientific reductionism is not only capable, but *sufficient* for dealing with the brain—an object often characterized as the most complex entity in the universe! Most science is reductionist: Our understanding depends upon reducing everything to its most basic components (atoms, elements, neurons, etc.) and building up a picture of how these parts interact when lots of them get together to form a system. In school, we learned about the very earliest machines, levers and wheels, and how complicated machines can be built up from the composition of components with known individual behavior. A system's behavior, then, can be deduced from the interactions of these components. Reductionism has enabled engineering and science to develop into two of humanity's crowning achievements.

But some systems are so extremely large, or so extremely complex, that they seem beyond the reach of reductive analysis. This includes all social systems, such as economics and finance. Human interactions are involved here, and this has consequences for social robots too, which we will explore further in Part III. The "machine" characterization tends to conjure up our early ideas of simple mechanisms, simplifying and trivializing the enormity of the idea and closing off other concepts.

Systems with lots of components, particularly those where the number of interactions between the components is high, may behave in ways and exhibit features that do not exist in any of the individual components. These are known as *emergent properties:* properties that are new and unexpected. The fields of complex systems theory and chaos theory explore the phenomena of emergence and the apparently random behavior often produced by complex, nonrandom machines.[9]

Complex system effects tend to occur when large systems contain nonlinearities, multiple feedback loops, fractal structures, or self-organizing internals. These are found in large engineering systems, all kinds of living cells, ecosystems, and societal systems like economic, financial, and business networks and organizations. And, of course, the brain, being a very large complex system with its 20 billion nonlinear neurons in the cerebral cortex alone and the feedback loops from its 2 trillion interconnections, belongs to this category as well. This is where the limits of reductionism become apparent.

The second, more important objection is that this view encourages technologists to concentrate on building a single artificial version of the brain: an *individual* brain. This entirely ignores the interactions among brains that form an absolutely crucial context for their behavior. In our early years, our comprehension, understanding, knowledge, skills, and other cognitive abilities (all growing at a fantastic rate) depend on interaction with at least one attendant caregiver. Without parental care, we do not develop and thrive. Then, in adult life, we form groups, societies, and organizations through which we survive.

As we saw in chapter 1, the societal aspects of human life suggest that intelligence is not bounded and contained within individuals, but rather exists across populations and is influenced by the culture of a society. On an individual level, the activities of the brain depend on the activities of the whole organism, the body; modern robotics has been helping to show how important this is. Embodiment, constructivism, and primacy of action are key principles that expound on how the life process, the life cycle of the individual, affects cognitive growth. We will examine these ideas further in Part II.

BASIC ARTIFICIAL NEURAL NETWORKS (ANNS)

ANNs have been around for more than 50 years. The first simple mathematical model of a neuron, created by McCulloch and Pitts in 1943, had a series of inputs which were added together, and if the sum total of the inputs was equal to (or greater than) a given value, then the model neuron would issue an output signal. These were known as *threshold neurons* because they fired if the input sum reached or exceeded the given threshold value. Each input to every neuron has a weight value attached (representing the strength of that particular connection), and these values are usually multiplied by their input to give a weighted input. Note that the weights can be either excitatory (positive) or inhibitory (negative). The fundamental operating process of ANNs is explained in box 9.1.

It was proved mathematically that these threshold devices are universal computing elements. That means that *any* logical or computational device can be built using only an unlimited supply of identical copies of just *one* of these artificial neurons. This universal component principle is also a key factor behind the success of modern digital technology because it also applies to the transistor logic gates that are the basis of practically all digital systems.[10]

Box 9.1
Basic ANNs

> The neuron model used in many ANNs involves a variable attached to each input, known as a *weight*. These represent the variable strength of the synaptic junctions in the brain, where one neuron makes contact with another. Figure 9.1 shows a very small network, with five neurons arranged in two layers. The weights are shown as black dots on the inputs, and the circles are the neuron bodies where the inputs are summed and tested against the threshold. The rectangle represents some input source; this could be an image, and a few selected pixels have been connected as inputs to the neurons in layer 1. If this looks a bit like the image processing scheme in chapter 6 (figure 6.5), that's because it is another way of doing the same thing!
>
> Let's suppose that we want an output signal from this network when it detects the shape in the input image (i.e., a kind of pattern recognition task). The pixels on the shape are giving strong signals (the four black pixels) and the background is giving no signal (the four white pixels). Now consider n_1; we want an output from this neuron *only* when *both* inputs are on. The input signals are always multiplied by the weights, and then added together; and if the total reaches the neuron threshold, it produces an output. So, assuming $0 =$ no signal and 1 is a strong signal, for n_1, the weights could both be 1 and the threshold could be 2; in that case, both inputs have to be active to make the neuron "fire." For the next neuron, n_2, we need an output *only* when *both* inputs are inactive; a suitable scheme would be both weights $= -1$ and the threshold $= 0$. Have a try at working out the weights for n_3 and n_4.
>
> For layer 2, we want n_5 to fire when *all* the other four neurons produce an output. This is an AND function (i.e., it gives an output only when neuron 1 AND 2 AND 3 AND 4 are active), so weight values of 1 for all four inputs, and a threshold of 4 will achieve this.

DIFFERENT APPROACHES: AI AND BRAIN SCIENCE

Simple models of neurons, such as in figure 9.1, are gross approximations that gloss over a great deal of biological detail and cannot claim to have a close relation to real brains. These artificial neurons give numeric outputs (sometimes only binary), but real neurons generate streams of pulses at different frequencies, and often in bursts. So more detailed models were developed (spiking neurons), in which some of the more detailed electrical behavior was captured.

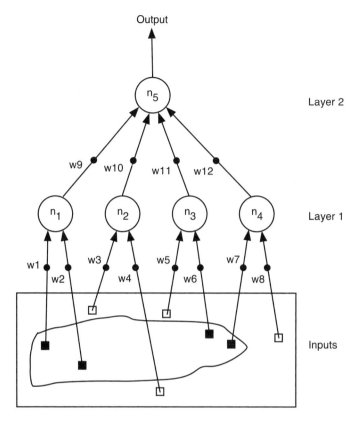

9.1 A simple ANN with two layers of neurons.

At an even higher level, the most authentic neuron models incorporate all kinds of chemical and hormonal effects that take place at the synaptic junctions between neural cells. These models are so sophisticated and have so many parameters that only a small number of neurons can be simulated on a typical desktop computer. This means that the level of biological authenticity that can be achieved often has to be balanced against the speed and size of the network that can be simulated. However, it is not always necessary to strive for maximum neuron-modeling authenticity at the same time as maximum network size and connectivity. It all depends upon the goals of the research.

As we have seen already, we can really judge an AI system only in terms of its purpose; and this is true for brain modeling too. Thus, highly detailed neuron models might be used in the simulation of a very specific piece of brain to explore a particular localized mechanism, or, on the other hand, very crude and simple neuron models could be sufficient in a huge network simulation that is exploring a large-scale connection hypothesis.[11] The role, purpose, or goal of the work determines how it should be judged. This important point cannot be stressed enough.

An example of this is the frequent failure of the media to recognize the two very different motivations for using ANNs: AI and computational neuroscience. If you see Gwen designing a network, writing code, and running computer experiments, you, most likely, would not be able to tell if she were an AI engineer or a computational neuroscientist. Both jobs look similar but have entirely different goals. Intelligence is only a minor concern for brain scientists. Computational neuroscience is a discipline that uses computational and mathematical models purely to further our understanding of the brain. Computational neuroscience builds models at different levels, from microneural chemical and electrical events through to network dynamics and architectural properties. The model being used must accurately match the biological phenomenon that is under study. In neuroscience, the model matters even more than the results—one could say that the model is part of the results; it is certainly part of the research contribution. Computational neuroscientists often use lots of computing power, mainly to run sophisticated software simulators.[12]

While biological authenticity is vital for computational neuroscience, it is not necessary in an ANN. Of course, information about the brain provides inspiration for new network design ideas, and so some of the properties of an AI network might relate to biological knowledge. But the AI laboratory is more concerned with results and specific applications; biological authenticity matters less than powerful learning methods. So, just because an AI model may look like a brain study doesn't mean that it is. And conversely, brain science isn't AI research, even if it looks similar.[13]

It would be nice to say that the flow of inspiration and ideas is bidirectional between AI and brain science, but it is actually massively one-way. AI picks up all the latest results and theories from neuroscience and, unconstrained by the biology, is able to experiment and innovate with the goal of producing new technology.

MORE ADVANCED NETWORKS

The layer arrangement seen in figure 9.1 is the simplest form of network. These *feedforward* networks (in which data flows forward from the inputs, always toward the outputs) are fine for pattern recognition and classification problems, but they have no memory, and therefore they cannot deal with anything where the output depends on the history of previous inputs.

Many cognitive functions involve patterns in the *sequence* of the inputs. Speech and text are obvious examples, where the sequencing (of sounds or symbols) carries much of the information. In fact, nearly all communication systems are sensitive to the specific *order* of groups of input symbols. This relates to episodic knowledge (as discussed in chapter 5), where the events occur in a narrative or recipe that entails meaning.

Recurrent neural networks (RNNs) are a special class of network that can learn and capture sequential data and episodic patterns. In addition to the normal feedforward layer structure, RNNs have feedback paths from the neurons in the near-output layers back to layers nearer the input. This allows circular excitation routes through the network so that it can be active even when there is no input. The output of the network depends upon both the inputs and the current state of the system. This allows the network to respond to long data sequences or events separated in time.

A particular area of success for RNNs has been in language processing. Recent research results for handwriting recognition and speech recognition have been impressive. Fully connected handwriting can now be machine-read in languages such as Chinese, French, and Arabic. There have been regular competitions for language-processing challenges, and as for vision, Deep Learning methods have been producing the best results. Language translation, textual query answering, and keyword spotting show excellent results with RNNs. Even music composition and rhythm analysis have been demonstrated by these methods. RNNs can also deal with long temporal gaps and noisy data and find use in protein analysis, stock market prediction, and many other sequence-based problems. See Schmidhuber (2015) for a review of deep neural networks, including their origins and background.

PREDICTIVE CODING AND AUTOENCODERS

Potentially, there are as many designs of network architecture as there are ways of connecting a pile of neurons (i.e., enormous!), but a number of distinct types with various features and characteristics have been grouped together under various names, such as *self-organizing networks*, *Hopfield nets*, and *associative nets*. We do not have the space to explore these here, but useful overviews are available.[14]

A recent idea, known as *predictive encoding*, looks very promising as a brain model and also has potential for sensorimotor control and learning in robotics. Instead of the simplistic input → process → output computing idea of perception, predictive coding assumes that the brain constantly predicts sensory input and compares the actual input with the expected input. This means that the brain has a model of the sensory signals and tries to keep this in step with the real thing. If the input is very close to the expectation, then the model is good and the comparison error is low. But if the input is different, then the comparison error will indicate how much the model needs updating. As the model responds to the errors, it self-improves and its predictions get better. Furthermore, the model can have expectations about the precision of the input so that confidence in predictions can be managed. For example, errors in prediction for vision in misty conditions would be expected to have lower precision than on a bright, sunny day.

This method has various attractive features. It efficiently maintains a model of the world that is constantly updated by sensory experience. It only deals with differences, so the whole sensory input does not have to be processed all the time, and the size of the error drives attention toward more interesting events.

Instead of the conventional bottom-up model of perception, predictive coding has a strong top-down influence on what is perceived. This not only explains the many feedback paths in the sensory processing regions of the brain, but also suggests how perceptual models of the sensed environment could be built up. ANNs that perform this kind of input error comparison are often known as *auto-encoders*, and they have been finding applications in robotics. See Clark (2013) for an overview discussion of predictive coding, and Muhammad and Spratling (2017) for a model of eye/head/body coordination on a simulated iCub robot.

Other artificial neural ideas include generative adversarial networks (GANs) which use two competing networks, and capsule neural networks (CapsNets), which improve hierarchical recognition performance.

ISSUES WITH ANNS

ANNs do not have to worry about accurately modeling the brain, but they do have some significant worries of their own. Some are technical and will eventually be solved, some are features of the method that limit the scope of their application, and others raise serious questions about their integrity.

Designing rather than programming: ANNs seem very attractive because they do not have to be programmed to perform tasks, but rather can learn for themselves. Unfortunately, the programming problem has been replaced by the design problem: How many neurons should be used in an internal layer? How many layers are needed? What connection pattern and computing functions should be used? There are methods for growing networks by automatically adding neurons,[15] but finding the right configuration of an ANN for a given problem is still a significant design issue.

The training cycle: Training on data usually requires very large data files and takes a long time. The data is repeatedly cycled through until the weights have settled down, indicating that the number of errors has dropped to an acceptable level. These training cycles can run into hundreds and can take hours or days to run. In the Google DeepMind evaluation function for the game of Go (as discussed in chapter 6), there were 29.4 million entries in the training set and 630 million weights to be adjusted.[16]

Catastrophic forgetting and knowledge transfer: A conventional multi-layer network will lose its original training if it is trained on a new set of data. During training on the new problem, the weights shift toward solving that problem and move away from their best values for the first problem. This is called *catastrophic forgetting*. The question of how to retain past knowledge is known as *transfer learning* and is an active research area.

Subsymbolic opacity: Anything and everything that an ANN has learned is encoded and distributed across the weights of the network; there are no words or readable text inside ANNs. In engineering, this is known as a *black box* because looking inside does not explain how it works; the system is opaque.[17] This is a big issue. In most AI work, even if words or text are not involved, there will be some means of observing how clusters are forming, categories are growing, or models are being populated. However, with ANNs, we don't get such help.

Implicit knowledge: Closely related to opacity is the question: What does our network actually "know"? It is a requirement of all human systems of judgment, estimation, and assessment that they give explanations for their decisions or actions. Any ruling, report, advice, or guidance can be challenged to explain the reasons for its recommendations; this is seen as a basic human expectation. So any learning systems must also explain themselves if they are to be accepted. Unfortunately, ANNs are seriously lacking in this respect. Not only is it very hard to construct an explanation for a result, but when an ANN produces errors, it is not easy to see why.[18] New visualization tools are being developed to better understand these complex networks (Zeiler and Fergus, 2014).

Verification: Safety-critical scenarios are where black boxes are in real trouble. Many computer-based applications do not endanger life and limb if they go wrong or produce errors. Faults on phones, unexpected downtimes, and failing domestic appliances are all annoying and inconvenient, but they usually do not kill people; they just waste time and money. But safety-critical systems are a class of application area in which any error (i.e., deviation from specified performance) has the potential to damage or even end human lives. The most obvious safety-critical areas are health and transportation. Medical systems can be very sophisticated, and they can directly influence treatment or control vital functions. Transportation covers cars, planes, and trains, as well as any machine in which people can be hurt or killed if a vehicle acts in unintended ways.

The process of verification is mandatory in the engineering of safety-critical systems. *Verification* refers to procedures intended to check that a product meets the design specifications, often expressed as follows: Has the product been built right? Addressing this question involves checking every possible state of the system against the safety criteria. This is an exhaustive process (not to mention exhausting). Just contemplating this for ANNs confirms that it is impracticable.[19]

Conservatism is a very powerful influence in safety-critical systems. In railway engineering and space flight, for example, old technology is used well past its expiration date, until the latest advances have been proven through extensive use. For example, in the first years of NASA's Space Shuttle flights in the 1980s, the shuttle used the old technology of magnetic-core memory because it was the most reliable nonvolatile memory at the time. It's difficult to see how ANNs can get over the verification hurdle for serious safety applications.[20]

This not only applies to hardware and software systems, but to AI learning in general. According to Rich Sutton's Verification Principle: "An AI system can create and maintain knowledge only to the extent that it can verify that knowledge itself."[21] In other words, AI systems should be able to tell for themselves whether they are producing correct results. This resonates with the idea that real intelligence involves the ability to explain what it sees, does, and understands. This is relatively easy to do for playing chess but is an unsolved problem for real-world knowledge. I don't mind my doctor giving me the story of the side effects of a treatment when they are proven estimates from a neural net, or a statistical system, but I would be much more concerned to learn that my intensive-care system, which is controlling my vital functions, is based only on a neural network that has learned from its mistakes!

It seems that ANNs get rid of a lot of programming compared with symbolic AI, but this work is often replaced by demanding initial setup, laborious training regimes, and lack of integrity through intrinsic opacity. As sophisticated software tools are increasingly produced to speed up and simplify the process of building these networks, attention will shift to the proper engineering of their role in ethical, economic, and critical applications.

SIMULATION PROBLEMS FOR ROBOTS

The proponents of both Artifical General Intelligence (AGI) and giant brain simulations often claim or imply that once these achievements have been realized, such systems can be easily connected to robot hardware to provide superintelligent robots. By making the assumption that robots are just input-output devices, they display a serious lack of understanding and illustrate the gulf between AI and robotics.

The concept of building a "brain" in a computer and connecting it to other devices is reminiscent of the "brain in a tank" scenario from philosophy. The idea is that some mad, Frankenstein-like scientist might remove a human brain and then keep it alive in a tank of nutrients with supplies of oxygenated blood and controlled environmental conditions. This brain is connected to external sensors and motor devices, such that it can operate these devices in the same way as it might do as if it were in an actual body. Note that this "body" could just as easily be a robotic hardware device or a software simulation.

The philosophical argument then goes: If the signals to and from the brain in a tank are exactly the same signals and responses as would be seen when the brain is in a real body, then the brain *could not know* whether it was actually in a body or connected to a simulated setup. Obviously, the simulation would have to be extremely authentic for this to work. This scenario is then reflected back to humans, and the philosophers' argument continues to say that we humans, therefore, also can't tell if we are composed of brains in bodies or are actually living in the simulation, just as in the film *The Matrix*.[22]

Of course, for most people, this is a ridiculous argument. The idea that every little movement of our fingers when we scratch our head, say, could be repeated exactly in a simulation, and in such a way as to generate and duplicate exactly the right feelings and responses that we feel when this happens, seems absurd. To achieve this kind of feedback to the brain would require an enormous amount of very detailed and complex modeling—so much so, in fact, that it is beyond any practical possibility.

The killer argument is that it would not only have to be extremely authentic for this to work, but that it would have to be totally authentic, with absolutely no errors. As soon as the slightest timing mismatch occurs, a glitch or delay in the simulation, it would be detected immediately as some kind of warp in reality. However, philosophers of the skeptical school insist that it is theoretically possible, and if we read their extensive and difficult justifications, we might be forced to agree that it is possible—*just* barely. But it is so extremely unlikely that the probability is very, very, near zero, and, thus, most people are happy to believe that the world we experience is the world that we live in.

It's important to note that I am not objecting to the use of connections to a separate supercomputer. If your robot does not have room in its body for a computer, then it's fine to use a radio link or other means to get access to the processing power needed. Indeed, so-called remote-brained robotics has been practiced for a long time.

The reason I raise this scenario is to bring home the weakness of current thinking that assumes (1) human beings can be easily "digitized" and simulated, and (2) the brain and body can be treated separately and differently.

You can't just connect a working AGI or brain model to a robot and expect any available brain knowledge or information to have meaning for the robot. If the computer brain has not been built simultaneously with the robot, then the data from the brain will have little meaning for the robot, and the robot's experiences will be alien to the brain. The only way for the connection to carry semantically appropriate messages, in both directions, is for the brain and robot to be treated as a single integrated system, and for the knowledge and expertise of that system to be developed at the same time.

OBSERVATIONS

- *Computational neuroscience* is the study of the brain by building computer models of parts of the brain. This often involves simulating various biological neural networks. The goal is to understand the workings of the brain, and it is not focused on intelligence.
- ANNs use very similar techniques, but designed for different ends. AI research isn't brain science, even if it looks similar. And computational neuroscience may look like AI research, but it isn't.
- The brain has always been a source of inspiration and ideas for AI. A current example is the promising concept of predictive coding.
- There are many kinds of ANNs. Widely different functions are obtainable from different architectures.
- Simple networks can only do simple things. Large, complex networks are needed for complex tasks. This fact may seem tautological, but it took a long time to establish!

- WBE is an extreme form of computational neuroscience, and it is fraught with difficulties, both technical and theoretical. These include the interpretation problem of understanding the internal workings of complex systems; the switching-on problem of setting up the starting state for a brain model; and the authentic input/output problem, referring to the need for very high-quality contextual embedding of brain models.
- Considerations of the brain and body show that they must be treated as parts of an integrated system, not treated separately and differently.
- Scientific reductionism and traditional machine models are not enough for understanding complex systems with many highly interacting components, such as whole brains, social systems, or human culture. Complexity theory and chaos theory are examples of tools needed to offer insights into such systems, although they won't be sufficient.
- The main challenges in traditional AI are finding good algorithms and clever programming. The first challenges in ANNs were in setting up an appropriate system: configuring suitable networks, arranging effective training, and understanding the system; but these are being overtaken by validation, integrity, and providing correctness measures for the results.
- The opacity of ANNs causes real problems for their application. Black box components are not allowed in safety-critical systems, and verification is a serious obstacle for ANNs. Even in less critical applications, explanations or justifications for results are required or expected.

10

BOLTING IT ALL TOGETHER

At the William Morris Institute of Automation Research, happy days are spent programming computers to carry out menial tasks such as writing newspapers, watching sports, and saying prayers.
—Cover blurb from Michael Frayn, *The Tin Men*, 1965

There has been remarkable progress in artificial intelligence (AI) in the last decade or so. It seems that many of the components that make up the wide range of skills and facilities that we call *intelligence* have been implemented in one form or another. Thus, we now have some excellent speech, text, and image-processing systems, as well as advanced learning and statistical reasoning methods. Masses of data are routinely collected, processed, and analyzed by AI learning systems. Robots now have good dynamic control of locomotion, and tactile and spatial sensing is improving. Novel plastics and other materials are helping produce cheaper, more user-friendly robots. This promising state of affairs leads to an obvious next thought: Can we put all these component subsystems together to make a complete, intelligent robot?

This is a very attractive proposition, and the idea has gained popularity as AI researchers look for new challenges. There have been conferences and workshops specifically focused on how this might be done and what approaches and methods are appropriate. Technical workshops call for papers "on the integration of AI techniques to achieve intelligent behavior" or "that integrate two usually distinct areas of research [AI and robotics]."[1] They also report that "the field of AI has fragmented into many challenging subfields that require and often reward isolation and specialization." And they bemoan the "lack of mainstream AI venues for publishing integrative research ... that combines techniques from multiple different fields to achieve a working robot system capable of complex behavior." The fragmentation into subfields is definitely an issue in AI, but I think the lack of publishing venues is less of a problem than the lack of integrative work to report.

This modular approach can claim some inspiration and plausibility from the modularity of the brain. It is known that different regions of the brain tend to perform different functions. For example, input from the eyes is processed in the occipital region of the cortex, at the back of the head. Meanwhile, in the temporal regions, there are motor control centers with adjacent areas that contain sensory inputs closely related to motor action. Higher-level functions such as abstract thinking and reasoning are found toward the top and forehead regions of the cortex. Other noncortical parts of the brain include the amygdala, where emotional states are influenced, the cerebellum, which deals with fine motor control, and the basal ganglia, which handles rapid attention selection and sensory prioritization. And so there really is some biological justification for thinking that a complete intelligent system could be made from a group of interconnected specialist modules.

This idea of modular brains has created a rich source of theoretical and psychological questions; for a psychological view, see Karmiloff-Smith (1995); for the philosophical angle, see Dennett (1998). However, we will focus on the engineering aspects of creating such systems. What are the difficulties with this building-block approach? Will it work? In fact, it turns out to be quite a bit harder than it might appear.

THE COMPLEXITY OF MODULAR INTERACTIONS

Let's consider the "bolting together" scenario for some existing robot technology. We now have humanoid robots that can walk over all kinds of surfaces, recover from slipping, or even being pushed or kicked, and stand up again if they do fall.[2] We also have robots from Yamaha that can ride an unmodified motorbike at speeds of up to 120 mph around a track,[3] and we have violin-playing robots from Toyota.[4]

So, an interested bystander might ask: Are we likely to see a single robot that can do all these things? The common feature across all three robots is that they all have to respond very rapidly to a multitude of constantly changing signals about the dynamic situation they are in. So, yes, it seems this could be feasible. A single robot would have to be trained or modified to combine all three tasks but it could, in principle, be based on one powerful dynamic control system that would achieve mastery of walking, riding a motorbike, and playing the violin.

Well then, our observer might say, we all know that computers are very small now and getting more energy efficient all the time, so we could put a system like Watson into our walking, bike riding, violin-playing robot. And then, could our robot also answer any questions using Wikipedia? Yes, that's true too; there seems no reason, in principle, why that can't be done. And next we might add the latest in face recognition functionality, so that the robot will take instructions from selected members of the family, and not from strangers or mischievous children.

There seems no reason why this process cannot continue, with more and more AI functions added to the robot system as desired. Current digital technology does not preclude any of this, and we can predict that such developments will happen. However, this incremental approach is not the same as integration. For example, our robot might be able to play the violin *and* answer questions about music or how violins are made, but it will not be able to talk about its own playing or explain anything about what it's like *being* a violinist. Similarly, it may recognize a family member *and* talk about faces and family trees in general, but it won't be able to empathize with the owner's particular family relationships. The missing element is the subjective understanding gained from direct and embodied experience. By having a "self," we are able to coordinate all these mental thought processes from a single viewpoint. Of course, our robot might try to pretend to have a personal life by looking up and using responses made by humans in similar situations. But this fake behavior will soon be detected, and trust will be lost. This shows again why the Turing test is so difficult.

This scenario exposes the barriers inadvertently created by the modular approach, which fragmented AI can hardly avoid. The main barrier is the difference in architectures, languages, and representations used by various modules. The motorbike robot can play an instrument because the tasks both involve the same skills: fine dynamic control. But that's no use at all for textual or image problems, where other, completely different systems must be recruited. AI has a history of *hybrid systems*, where two radically different AI systems are connected together in some way to combine the advantages of both. For example, by connecting a neural network to a symbolic reasoning system, we might gain the learning ability of a network, together with the power of logical reasoning. Let's assume that the aim is to both learn some entities (say, features in images) and also reason about them; the simplest approach would be to let the two separate systems communicate through some kind of link. So the image engine might pass messages to the reasoning engine (well-contained computational units are often called *engines*) along these lines: "I figure there is a house and six children in this image, what does this relate to?" The reasoning engine might reply: "Previous similar cases have mainly been about schools. See if there are any features in the image that correlate with features in images of schools."

There are two things to notice in this setup. First, the connection is not just a matter of plugging the systems together; it requires some form of translation scheme. Of course, they can't communicate in English: They don't have *any* language. They don't even have common symbols for the same things. Words are meaningless to the image engine, just as images are to the reasoner. So the images used will most likely be tagged with labels indicating their content; in this way, the reasoner can use words as its basic data. But who is going to do the labeling? In most research, the tagging of images has been done by hand because there is no way to automate this subtle, all-to-human process. This is a major weakness. Some algorithms have been developed to link visual features with social labels and other signals from social platforms on smartphones, but this is still labeling by human activity.

The second issue is that the learning and reasoning engines are still separate and work on their own. Yes, they cooperate, and this does lead to enhanced performance, but the integration is at the weakest level—the learner can only learn about fragments of images and improve its recognition abilities, and the reasoner can only reason about the labels that the learner provides. A more comprehensive solution would facilitate reasoning during learning within the ANN engine, and learning during reasoning within the reasoning engine. This would be a proper merger of the two capabilities, allowing deeper learning and improved reasoning.

Most research on hybrids has involved combining only two systems. If there are more than 2 modules, then the number of possible connections increases considerably; 10 modules require 45 separate connections. The way a round this is to use a "command center" model, where a central interchange point connects to all the other modules.[5] For example, there are 24 official languages in the European Union, and to translate all of them would require 276 individual translation services; but if one language is designated as a core language, into which all the others are to be translated, then only 23 translation services are required, or 45 for two core languages.

HOW CAN COMPUTERS REPRESENT WHAT THEY KNOW AND EXPERIENCE?

The difficulties outlined up to now all hinge on the one of the most profound AI questions: How should knowledge and experience be represented? The basic materials of thinking—concepts, facts, perceptions, experiences, and so on—somehow have to be accessible and be represented in some form. We don't know how the brain stores and processes experience, knowledge, and memories, so various arrangements have been tried for computers. For example, arrays of pixels are used for images, symbols capture textual content, and numeric variables are used to express the strength of a belief or the reliability of a fact. AI representations are carefully chosen, and indeed tuned, for their respective tasks, and many kinds exist for different purposes. Consider a very simple example: the concept of a box. What do we need to *know* about a box in order to understand boxes and make sense of them? Table 10.1 illustrates some possible box elements. Some of the properties are essentially physical facts (about objects), while others concern possible behaviors or actions. Some, like

Table 10.1 Characteristics of the concept *Box* that may need representation in intelligent systems.

A BOX		
Property	**Type**	**Examples**
Physical	Texture, strength, mass	Corrugated, soft, weight
	Materials	Cardboard, plastic, metal
Size	Spatial dimensions	3.4, 5.0, 7.8 ......
Shape	Cube, cylinder, ...	2D or 3D description
Appearance	Viewpoint, perspective, scale	Occlusions, transforms (e.g., rotation)
Usage	Containment	Container, receptacle
	Support	Stand, height increase
Function	Storage, transport, support	Contents, route, location
Operation	Stacking, filling, emptying	Accessing, retrieving, moving
Behavior	Structural change	Open, close, collapse (cardboard)
Names	Textual	Toolbox, letter box, mailbox, waste box, signal box, ballot box, black box, strong box, nest box

Usage and Function, are human-assigned meanings, whereas Operation refers to things that can be done to a box. In fact, there are no guidelines for this because there are no set of fixed attributes that cover "boxiness"—there are endless possibilities depending upon what experience of boxes is being represented. It all depends on what the knowledge is to be used for—once again, it is the task that matters.

The idea of a box is a relatively simple concept. Notice that we have not included metaphorical cases, such as: "He's getting on his soapbox again!" or "It checks all the boxes." Many of the words that we use relate to concepts that are much harder to define because they carry much more complex meanings. And the usage of words changes; words can have multiple meanings or hidden meanings.

These examples illustrate why the integration of various AI modules unavoidably involves reconciling the different representations that they use. Representation has always been a hard problem area in AI and has become a research field in itself. It started with quite wide interests across language, logic, and reasoning methods, but now also has a Deep Learning community. For example, the International Conference on Learning Representations is predominantly concerned with ANNs, and in 2018 it attracted some 500 papers and over 1,000 delegates.[6]

THE LIMITATIONS OF TASK-BASED AI

The task orientation of AI modules becomes a problem when we try to reconcile them into an integrated system. If we know the set of tasks we wish to combine, then we can select the modules and attempt the difficult challenge of combining them. But a complete, intelligent robot that can deal with all the variety and complexities of life does not equate to a task. We want an autonomous agent that can learn from its mistakes, develop new skills, and adapt to changing circumstances, but how do we design one when we have no idea of what those mistakes will be and how those circumstances will change? The most notable feature of a real general-purpose robot will be its supratask abilities.

In chapter 8, the idea of general intelligence was introduced as "a non-task-specific but general capability that can be applied in whatever circumstances arise." If realized, such a system should be able to perform practically all AI tasks and would threaten to make conventional AI obsolete.[7] This rather nebulous concept would be ideal for a general-purpose robot, but is it really attainable?

GENERAL AI

General-purpose intelligence has always been the "holy grail" of AI, but unfortunately, it has proved entirely elusive. Most direct investigations of Artificial General Intelligence (AGI) are theoretical, involving formal, mathematical, and philosophical studies (often involving hard math!). The results usually concern the limits of computability and the complexity and feasibility of possible future algorithms. Such theoretical work has searched for powerful, general methods that will deliver precise and accurate solutions to intelligence problems. Usually, the problems used in this work concern prediction, on the basis that powerful prediction abilities are a key skill of any intelligent system. Interestingly, several theoretical limitations have been discovered. It was found that simple general algorithms cannot exist and any powerful computable predictors will be necessarily complex (Legg, 2008). Even worse, very general prediction algorithms cannot be found mathematically and must be explored through "experimental endeavour due to the difficulty of working with the required theory" (Legg 2008, 105).[8]

The focus on prediction may be mistaken. David Deutsch argues strongly against the conventional wisdom that intelligence relies on predicting and extrapolating from known data: "in reality, only a tiny component of thinking is about prediction at all."[9] He rejects the widely used inductive approach, whereby thinking is seen as a process of predicting future patterns of sensory experience from past ones, and claims that what is needed is a new epistemological theory that explains how brains create explanatory knowledge. Being able to explain and create explanations is what makes us human, says Deutsch, who sees a Popperian process of conjecture and criticism as the way forward.[10]

Such a theory does not yet exist, but Deutsch puts his finger on the issue when he asserts that AGI will be qualitatively, not quantitatively, different from all other computer programs: "It was a failure to recognise that what distinguishes human brains from all other physical systems is qualitatively different from all other functionalities, and cannot be specified in the way that all other attributes of computer programs can be. It [AGI] cannot be programmed by any of the techniques that suffice for writing any other type of program. Nor can it be achieved merely by improving their performance at tasks that they currently do perform, no matter by how much."[11] So task specifications and other normal engineering methods will not be sufficient.

MASTER ALGORITHMS

At a less theoretical level, some people are searching for a "Master Algorithm" that combines all the machine-learning (ML) techniques into one single method. One proponent of this concept is Pedro Domingos, who hypothesizes that: "All knowledge—past, present, and future—can be derived from data by a single, universal learning algorithm" (Domingos 2015, 25). By analogy with physics, where a single equation (an abstraction) can cover all realizations of a phenomenon, Domingos believes that a similar kind of abstraction can capture all learning. "The Master Algorithm is the unifier of machine learning: it lets any application use any learner, by abstracting the learners into a common form that is all the application needs to know" (Domingos 2015, 237). Domingos is trying to combine five main ML paradigms: symbolic logic, Bayesian probability, neural networks, classifiers, and genetic programming. He argues at length for the idea, but at the end of his book, admits that, so far, we only have "a glimpse of what [the algorithm] might look like" (Domingos 2015, 292).

I find many difficulties with this approach. For a start, Domingos does not actually define "knowledge," but he describes the extraction of knowledge from data; by this, he clearly means textual, image, and related forms from large data sources. He clearly doesn't think human experiences or thoughts that are not available as data count as knowledge. While it may well be possible to derive all the world's *factual* knowledge from new Big Data sources (that statement would have sounded extraordinary just a few years ago!) such a result does *not* amount to an autonomous AGI. Yes, the algorithm (if it can be created) would discover useful things and produce more knowledge, but it would not understand much of the knowledge it is processing. Domingos (2015, 25) says, "Give it a video stream and it will learn to see. Give it a library and it learns to read." I think this reveals a shallow level of thinking displayed by some of the deep/Big Data learning enthusiasts: Learning is seen as the extraction of yet more data from existing data. If only it were that easy!

The mistake made by the Master Algorithm enthusiasts is staying inside the ML framework. By combining lots of ML methods into one grand supermethod, you still end up with another ML technique. It's another reliance on reductionism that doesn't serve us well. Video streams and libraries contain subtle, subjective human knowledge: knowledge that is not easily extracted as features or words, but which forms complex relations between events and can be discerned by other subjective agents. The early, simplistic computational notion: *input* $\rightarrow$ *process* $\rightarrow$ *output* produced a now abandoned metaphor for cognition: *sense* $\rightarrow$ *think* $\rightarrow$ *act*. AI is still moving away from this limited framework. Thinking is a complex process involving actions and perceptions in a highly integrated interplay. While we can learn much from data, just feeding in streams of data is very far from what biological systems need in order to learn and grow mentally. In chapter 11, we will see why much more is needed and how it can be obtained.

BIOLOGICAL COMPARISONS

The characterization of AGI is often seen as a model of human intelligence. Many people appear to assume that AGI, being the ultimate goal of AI research, is the same as achieving a working model of human intelligence. However, is this really true?

The pure intellectual power sought for in AGI is not typical of human intelligence. AI systems are always expected to succeed; progress is measured in terms of success rates, challenge performance, and task achievement. On the other hand, failure is characteristic of humans; we often try things that don't work out and are prepared to learn, adapt and make new attempts, or recover in other ways if we fail. Humans fail many times every day; usually these failures have minor or trivial consequences and are not even noticed. How often do you start something in the wrong order, correct yourself, and carry on? This is indicative of a *very* different cognitive approach. For anyone who doubts this, one only has to read some psychology that exposes the odd ways that humans *actually* think, such as the excellent *Thinking, Fast and Slow* (Kahneman 2012), *Irrationality: The Enemy Within* (Sutherland 1992), or *Predictably Irrational* (Ariely 2008).

It is not surprising that AI is compared with human intelligence; there is nothing else to compare it with. AI progress is measured in a series of milestones that have been reached or exceeded: master-level chess or Go, face recognition benchmarks, speech recognition challenges, navigating a passage (on the roads, sea, air, in buildings, or in the wild). These milestones and benchmarks are related to human tasks, and so we can often say that human-level performance has been reached, or even exceeded. However, it is very important to notice a subtle double meaning in the phrase *human-level AI*. All the breakthroughs seen so far are by AI systems that are specifically designed for their single task; each breach of best human performance is by a different system. So all the human-level AI in existence has been achieved by a motley collection of different AI systems. However, some use the phrase to mean a *single* AI system to achieve human levels on a wide range of tasks that require intelligence. For example, the annual conference series on AGI (the 12th was in 2019)[12] was held under the banner of "Human-Level AI," meaning "the creation of thinking machines with general intelligence at the human level and possibly beyond." This is just another way of describing AGI, but it is unfortunate, as it introduces confusion. Another variant on this is the term *Human-Level Machine Intelligence (HLMI)*, again meaning nothing more or less than AGI.

Nevertheless, neither AGI nor HLMI are the same as reproducing human intelligence. Human-level *performance* can be measured in terms of sets of target criteria, but achieving artificial human intelligent *behavior* is quite another matter. To measure progress in modeling or copying human intelligence requires holistic measures of activity, not just end goals. Human intelligence is not only determined by tasks, but by developmental histories, social interactions, environmental situations, and genetic predispositions. It is a complex phenomenon involving many cognitive features and social contexts.

This distinction is important. I don't like introducing new terminology, but it looks like we really need a new term here. Let's try to remove any vagueness in these long-term aims by defining *Artificial Human Intelligence (AHI)* as the endeavour to achieve *humanlike* intelligence, as distinct from AGI, which aims at very advanced general AI performance. AGI can use all the techniques of AI and ANNs, and indeed any other discipline, to reach its goal, while AHI must not only demonstrate human capabilities, but also must endeavor to do so in the same way as humans perform. Box 10.1 summarizes these various distinctions.

SUPERINTELLIGENCE (SI)

Recently, there has been much excitement in the media about the possibility of computers becoming superintelligent and overtaking human thinking power. *Superintelligence (SI)* is yet another term thrown into the pool of "very intelligent" terminology. We can take it to mean an AGI system that not only exceeds most, if not all, known benchmarks for human intelligence, but also continuously self-improves. The idea is that computers or robots would become capable of recursive self-improvement (redesigning themselves) or of designing and building new computers or robots better than themselves. Repetitions of this cycle might result in a runaway effect—an intelligence explosion—where smart machines

Box 10.1
Kinds of AI

> **Task-based AI** refers to AI systems that have been built to perform some particular intelligent function. All existing AI software is task-based.
>
> **AGI** is the property of an AI system that can perform most, if not all, known intelligent functions. This general-purpose AI is a theoretical ideal, and no practical examples have yet been realized.
>
> **Human-Level Artificial Intelligence (HLAI)**, is another way of describing AGI. Many examples of human-level performance have been achieved by specific AI programs, but HLAI requires generalization over all tasks by one AI entity.
>
> **Superintelligence (SI)** is a purely theoretical concept. SI refers to a level of AI performance that considerably exceeds all known measures of intelligence. SI is a form of AGI that could continuously self-improve, and therefore would subsume AGI or HLAI.
>
> **AHI** is the achievement of *humanlike* behavior and performance in some aspects of human intelligence. Drawing on both psychology and neuroscience, it is not as ambitious as some of these other options, but offers an alternative approach toward general intelligence for robotics.

design successive generations of increasingly powerful minds, creating intelligence far exceeding human intellectual capacity and control.

Because the capabilities of such SI may be impossible for a human to comprehend, the explosion event is an occurrence beyond which the implications for the future are unpredictable, and possibly unfavorable. This raises the possibility of circumstances where humans become subservient to a superior species of machine intelligence that could maintain themselves (through repair), evolve (through design and manufacturing), and ultimately manage their own survival. The key concern is that this "machine species" would develop and favor its own priorities rather than those of the human race. It might somehow gain control over the planet's network of computers so that energy and environmental parameters become attuned to machines rather than humanity. It might then get along without us, leading to the ultimate demise of *Homo sapiens*. This situation has been called a *singularity*—that is, a severe and sudden situation where the normal rules break down and we enter a new state or world order. This concept has become so popular that until recently there was a Singularity Institute that ran regular "Singularity Summits." These summits are now run by an educational outfit, and the institute has been renamed as the Machine Intelligence Research Institute (MIRI). MIRI's mission is to ensure that the creation of smarter-than-human intelligence has a positive impact: "We aim to make advanced intelligent machines behave as we intend, even in the absence of immediate human supervision."[13]

Of course, the ever-increasing power of technological devices that affect society raises justifiable concerns and feeds into worries about AI. Quite apart from the media speculation, there are serious academic studies underpinning these ideas. An example of this is the book *Superintelligence* (Bostrom, 2014), written by the director of the grand-sounding Future of Humanity Institute. This institute, at Oxford University, has impeccable credentials and works closely with the Centre for the Study of Existential Risk at Cambridge University, which has included such eminent scientists as Lord Martin Rees and the late Stephen Hawking, who both agreed with Bostrom.

Like many of the ideas we have explored in this book, the concept of a machine intelligence singularity is not new. In 1965, the mathematician I. J. (Jack) Good introduced the idea of an "ultraintelligent machine" that "can far surpass all the intellectual activities of any man, however clever." This machine could then design even better machines than itself, and this would lead to an "intelligence explosion," and the ultraintelligent machine would be "the last invention that man need ever make." (Good, 1965, 33). Interestingly, Jack Good worked with Alan Turing, both at Bletchley Park during World War II and afterward at Manchester University. Good also advised Stanley Kubrick, the director of the film *2001: A Space Odyssey*, on advanced computers, and so, quite possibly, HAL was based on Good's ultraintelligent machine.

The problem with all the proponents of SI is that they only offer a philosophical idea; it's all hypothetical, so there are no details as to how SI might work or how it might come about. Computers frequently get out of control, and that is often the case in AI work too, but it is hard to see how humans would not have the means to recover from this. And, as for other products, engineers would be expected to retain various controls over the autonomy they create in such systems. Topics like recursive self-improvement have been mentioned, but computational theory is a complex subject, and the authors need to connect the science fiction to reasonable technical scenarios.

INTEGRATING DEEP ARTIFICIAL NEURAL NETWORKS (ANNs)

The neural network research community are also bolting things together in their own way. In particular, the Deep Learning groups are aiming at superpowerful intelligent systems. For example, Google DeepMind says, "We combine the best techniques from machine learning and systems neuroscience to build powerful, general-purpose learning algorithms."[14] Some methodological ambiguity crops up again here; Google DeepMind is clearly aiming at the ultimate goal of AI research and views that as "Solving Intelligence," its slogan. Because of the phrase *combining of techniques*, I take this to mean that they are aiming at AGI as the ultimate goal rather than solving the problem of understanding human intelligence.

The various commercial Deep Learning laboratories are somewhat secretive, but we can read between the lines of their publications and get a good idea of the general agenda. Their ambitions including finding solutions for many of the issues that we have looked at so far; indeed they are going to attempt to create neural versions of most, if not all, of the AI problem areas. The exceptional results achieved by deep artificial neural networks (ANNs) in the various AI competitions and challenges have often involved training (i.e., supervised learning) on very large data sets. This has been a recognized weakness, and unsupervised learning is the current focus.

The AlphaZero story, discussed in chapter 7, shows the way things are going. Previous (trained) systems tackled the task of *playing* a game, and they ranged from human-tuned tree search to neural networks trained on massive game data. But we now have systems whose task is to learn *how to play* games from scratch.[15] The elimination of any supervision or training data is the significant advance. Of course, Big Data will continue to be a resource for learning, but the shift is away from being shown examples of desired behavior and toward learning and discovering the best behavior through independent experience. Such systems can now be found as softbots scanning the internet for new insights into user behavior, or in speech and image assistants that continually learn new patterns.

The deep-layered vision systems introduced in chapter 6 are being extended with reinforcement learning and feedback paths to form new recurrent neural networks (RNNs) that will selectively shift their attention to salient features in the image input. This approach is also planned for text processing: "We expect systems that use RNNs to understand sentences or whole documents will become much better when they learn strategies for selectively attending to one part at a time" (LeCun, Bengio, and Hinton, 2015). This shift away from the somewhat mechanical mass-processing of many possibilities and toward attending to particular problems in more detail is likely to lead to a more humanlike model of agency.

The vexed problem of representation is being addressed by new work in the field of representation learning. Instead of designing a structure to model what is thought to be the key features in some data, the idea is to automatically explore the data first and learn a suitable representation for best capturing what is to be learned (Bengio, Courville, and Vincent, 2013). This is important work because, quite apart from the implications for ANNs, by processing Big Data sets, we may gain a better understanding of the information that matters, not only as features for computers (harking back to Samuel's major issue of not handcrafting meaningful task criteria but getting the system to find those itself), but also relating to the human meaning in messages and other content.

There is another major concern: "[N]ew paradigms are needed to replace rule-based manipulation of symbolic expressions by operations on large vectors" (LeCun et al., 2015). This refers to the previously mentioned incompatibility between symbolic representations, used for textual data, and subsymbolic representations, as used for vision and speech. Although symbolic *data* is often used in ANNs, the symbolic *processing* techniques developed in AI for reasoning and inference have not (yet) been efficiently implemented in ANNs. Researchers in this area argue that "major progress in AI will come about through systems that combine representation learning with complex reasoning." This is already being explored, (e.g., Bottou 2014), and it is likely that the aim, at least from the

ANN community, will be toward *all* representations being subsymbolic, or at least compatible with deep neural models.

These ambitions amount to nothing less than reimplementing all the branches of AI into neural-based systems. All AI techniques could then be used to solve problems in neural networks. Getting all of AI into one formalism will make the integration issue much less challenging. If all modules can use very similar representation structures, the translation problem becomes much easier—maybe almost trivial. We can expect to see systems emerge that combine the low-level speech and vision achievements with similar advances in representation learning and complex reasoning. This must be the best bet at getting anywhere near integration through the bolting-together approach.

So, we can expect many benchmark challenges will be met in the near future, and they will often go well beyond human levels of performance. However, these advanced systems will still be lacking in two areas: (1) they will not be general purpose, and (2) they will not be humanlike. Although learning will be a major characteristic and they will have a range of impressive cognitive abilities, they will be designed for a purpose, there will always be a designer with some function in mind, and these aspects will constrain their ultimate capabilities. In addition, the operation and impenetrability of ANNs mean that the intelligence produced will not be humanlike; it will appear to us as alien intelligence. This rules out good-quality social robotics. Of course, some limited simulation of social interactions will be produced, but the test of real social cooperation will not be passed by this route. Last but not least, such systems will be subject to all the ANN integrity issues outlined in chapter 9, which imposes a real barrier for safety-critical applications.

OBSERVATIONS FOR PART I

In the broad review of the basics of AI that has been given in Part I of this book, we can see that conventional, task-based AI is inadequate for building the kind of general-purpose adaptive systems that characterize human behavior. AGI would offer the potential to fill the gap but, despite recently increased popularity and attention, there has been no concrete progress of any significance. Various road maps have been designed, and methodologies outlined, but no technical advances are apparent. The difficulties of AGI are well summarized by OpenCog, an AI research company that is pursuing a practical coding approach to AGI. OpenCog argues that (1) intelligence depends on certain high-level structures, (2) no one has discovered any one algorithm or approach capable of yielding these structures, and (3) achieving the emergence of these structures within a system formed by integrating a number of different AI algorithms requires careful attention to the manner in which they are integrated.[16] It states that, "so far, the integration has not been done in the correct way, which is difficult but not insurmountably so."[17] We have to wish the company well in finding the correct way to integrate, but the odds do not look promising.

We can conclude that the social robot, requiring humanlike behavior as well as social skills, only makes sense in terms of AHI. We need to reject the traditional AI design process and focus on human behavior and open-ended, general-purpose learning. This will need some form of internal problem-solving motivation rather than task specifications. This is explored in Part II of this book, which shows how to build developing robots using models inspired by the human sciences.

Let's list the main points developed so far:

- **AI has made enormous recent progress and is becoming ubiquitous but it will remain task-based.**

  Recent research has established modern AI techniques and, supported by vast data sources, new applications and services will extend into many areas of life. We can expect fields like medicine, health care, drug development, office automation, agriculture, and food production to benefit, not to mention consumer goods and social markets and services. As with all technology, there will be uses and abuses, but social, moral, and ethical forces must come into play to control and moderate their application. Regulation by governments will be important.

  AI systems will continue to be designed for specific purposes; that is, they will be task-based, with very little flexibility outside their particular application.

- **ANNs are advancing rapidly but have a number of serious issues.**

  The development of ANNs is set to continue and the dominance of this technology poses a threat to traditional AI methods. A significant goal for future research is the integration of different AI competences through the common substrate of neural processing. If this is achieved, it will produce another wave of enhanced AI performance. This approach may well succeed in producing robots that have a range of useful applications, but they will still be task-based and lack humanlike subjectivity.

A big problem with systems that learn is that they are always changing, which makes them very difficult to verify against standards of performance. An additional concern with deep neural systems is that they are functionally opaque and unable to give explanations or justifications for their results. These serious black-box issues do not engender trust, which means that they cannot be used where reliability is a top priority.

- **Doom and disaster do not follow from machine victories over humans. Arbitrary milestones are not the same as scientific breakthroughs.**

  Dire predictions of dystopic futures are made whenever AI outperforms humans at some task, but these negative outcomes do not materialize. It is research breakthroughs that change technology, not reaching arbitrary goals in competitions.

  We can expect future AI to go beyond human levels of performance in many other areas, with the result that we will learn more about the problem domain. There is plenty of evidence that society adapts and uses its own intelligence to embrace and assimilate most new technologies.

  AI and robotics are not malign in themselves; they are neutral and do not have an agenda. If any technology-led dystopia does come about, it will be caused by humans and will be an entirely human responsibility. This is because AI and robotics are still very limited in scope and range when compared with humanity.

- **There are fundamental differences between machines and people.**

  The things that are important and common to all humans (life, death, sex, birth, health, friends, family, children, shelter, food, sleep, etc.) are *all* entirely meaningless to digital systems. Machines may come to have knowledge of the basics of human life and death, but without

being able to experience these essential human characteristics, they will not have a subjective view that resonates with humans. This would be the same for us when meeting alien life forms, or other animal species to a lesser extent.

These distinctions raise two notable issues:

- **Unlike humans, AI and robots are not general-purpose agents.**
  Humans are generalists (i.e., general-purpose agents). They learn to deal with different situations, learn many new tasks, and adapt their existing skills to cope with all kinds of problems. If robots are to be general purpose too, then they will need to learn in a similar way. They must build their own subjective cognitive models from experience, rather than use objective knowledge from a designer or data source. This requires an open-ended general intelligence, unconstrained by a designer's requirements.

  Theoretically, AGI is required for such robots. But the "holy grail" of AI has comprehensively failed to materialize after 60 years of AI research. AI is still fragmented into separate task areas, and this looks likely to continue.

- **Brains need a body for full intelligence.**
  Bodies are necessary for learning through personal experience and for maintaining a subjective sense of self. Social interactions with other embodied agents are also a vital part of human intelligence. As Antonio Damasio's remarkable book argues, we are bodies that think—the body feels, the mind thinks, and both are essential and profoundly intertwined parts of human intelligence (Damasio, 2000). Thinking is diminished if you don't have a body because the body is not just an input-output device, but rather is the source of all experience that shapes the cognitive processes.

  This means disembodied AI and intelligent robotics are not the same endeavor—robotics is not a branch of AI, and we need to recognize the differences. Robotics has to deal with the integration of mind and body. By being disembodied, AI gains huge benefits of wide and flexible application, but it also suffers from having only second-hand experience of the physical world. Grounding and semantics are problematic for AI, but less so for humans and robots.

These issues lead us to the idea of social robots as a key challenge area for new research.

- **The social robot is an exemplar concept for examining subjectivity and the self.**
  Given that intelligence often involves more than the workings of an individual mind, social intelligence is a fundamental and promising area of AI. However, this is also probably the most difficult area because it involves general-purpose knowledge and understanding.

  The acid test for robots that think and behave like humans must involve long-term autonomy and free and meaningful interactions with humans. Such robots will be friendly and engaging, and be accepted by humans as social agents. Social robots will need to interpret human behavior and emotional states, engage with norms of gesture and idiom, and maintain mental models of their human contacts as individuals. This requires such robots to have an understanding of agents as distinct entities, including themselves. They need a model of "self" and similar models for "others." Self-models provide a subjective perspective and are necessary for empathetic understanding.

  Social systems face open-ended tasks, so they need general intelligence and a shift from designed learning algorithms to self-discovery. Real-life social interactions are based on personal experience and are deeply grounded in the subjective life history of the individual agent. All individuals have different personal experiences that have to be learned and cannot be generalized.

- **The field of Developmental Robotics offers a way of exploring the growth of social intelligence.**
  This is a new, goal-free approach, using mechanisms like play, for exploring open-ended learning, and it is a novel step toward general-purpose skill acquisition. It deals with the need for background knowledge, general problem-solving, and subjective viewpoints through experiential learning rather than data-driven approaches. This is the topic of Part II.

Finally, we can predict that the most intensively researched topics of the future will be in three areas. The integration of different deep neural systems has already been mentioned. The other neural challenge area is the safety-critical problem. If powerful neural network methods are to reach their potential in safety-critical applications, they will have to achieve the 100 percent elimination of errors that such vital tasks demand. But the issues of transparency for decisions and repeatable performance with high accuracy and reliability are major barriers.

The third and very promising area is combined human-computer systems. Many areas of AI are now recognizing that human-computer cooperation gives better, more useful, and more acceptable results than AI systems on their own. This has been known for some time in the manufacturing and production industries, and the approach is being actively promoted in many new application fields. It is obvious that people prefer human contact in medical situations, for example, but it is being realized that the results obtained from cooperating human-machine systems give better results (often much better) that either on their own.

II
ROBOTS THAT GROW AND DEVELOP

11

GROUNDWORK—SYNTHESIS, GROUNDING, AND AUTHENTICITY

The principal idea of enaction is that a cognitive system develops its own understanding of the world around it through its interactions with the environment. Thus, enaction entails that the cognitive system operates autonomously and that it generates its own models of how the world works.
—David Vernon (2010, 89)

It is going to be very difficult to build intelligent robots that think and behave like humans. At least, that appears to be the message from the review of artificial intelligence (AI) in part I. We explored whether robots can be as useful, intelligent, physically competent, and mentally flexible as people are, and we found evidence that very useful and intelligent robots will become commonplace in the near future, and many of them will supersede our physical and mental abilities in various ways. It is clear that both AI and robotic technology are currently on an accelerating curve, and some very powerful systems will be created in the next few decades. But they won't be as mentally flexible as people, and they won't think and behave like humans. They certainly don't at the moment, and there is little evidence to show that they are likely to do so in the future.

Future robots will still be obviously "robotic," and no one will confuse them with human beings; they will be engineered and built to serve various specific purposes. In part I, we reasoned that engineering methods do not provide the right approach for building humanlike robots. Of course, good engineering is vital, as it is the heart of technology and the basis of progress; but it is not sufficient to master this particular challenge on its own. We can't just engineer our way toward humanlike robots—we need something more.

What is needed is a different approach. But how can we have a different approach from engineering? After all, building robots necessarily involves hardware and software, both of which depend heavily on the engineering disciplines. The answer concerns the task-based approach, which I have argued underlies nearly all current research in robotics and AI. We need to shift our attention away from specific tasks, and toward trying to achieve desirable behaviors. Humans display their intelligence via their behavior, through their flexible lines of reasoning, through their adaptive ability to cope with change and new experiences, and in their general versatility.

You might ask: Why this obsession with human behavior—why does a robot have to be humanlike? There are several possible answers. First, it is generally agreed that humans are the prime exemplar of intelligence; this is why many scientific papers acknowledge that human intelligence is the ultimate target for AI. Second, human cognition uses many effective techniques that are very different from the expertise of machines and algorithms; this strongly suggests that new, interesting, and significant methods are still waiting to be discovered. Finally, there is a great deal of interest among psychologists, philosophers, and scientists in general (and the public) in discovering whether the brain is a machine that can be simulated in some way on a computer. By modeling human cognitive phenomena, we can learn about this issue in many ways.

But there is also another reason. We can state that reason in terms of a bold hypothesis (one of several that will be discussed in part II):

Hypothesis 1 *Artificial General Intelligence (AGI) has comprehensively failed to materialize following decades of AI research; Artificial Human Intelligence (AHI) is likely to be the only route to achieving any form of AGI.*

AI has clearly failed to make any advances in AGI; it may well be that AGI is an abstract concept that cannot be implemented. Perhaps the idea of a pure intelligence as an almost mathematical entity (a "brain-in-a-box") is just that: an idea. Certainly there appear to be severe theoretical limitations to finding simple, efficient, and general algorithms for intelligence.[1] Perhaps general intelligence exists only in autonomous systems with needs and wants. In any case, leaving aside the philosophical questions, the primary argument in this book is that approaching intelligence along human lines (i.e., AHI) is an important endeavor. It is a vastly underexplored field, and it offers real promise for addressing some of the outstanding problems in AI and robotics. It may also help to shed light on the possible mechanisms for human cognitive behavior. That is well worth doing!

THE CLASSICAL CYBERNETICS MOVEMENT

Now we've got the notion of a machine with an underspecified goal,
the system that evolves. This is a new notion, nothing like the
notion of machines that was current in the Industrial Revolution,
absolutely nothing like it.
—Gordon Pask, quoted in Haque (2007, 54)

There have always been individuals and groups interested in nonmainstream approaches, and that includes work sympathetic to our hypothesis. A relevant early research theme was known as *cybernetics*. The term *cybernetics* was introduced in 1948 by Norbert Wiener, a prodigious mathematician at the Massachusetts Institute of Technology (MIT). It is usually defined (using the subtitle of Wiener's book) as the theory of "control and communication in the animal and the machine" (Wiener, 1961). In Britain, Kenneth Craik had also been laying the foundations of cybernetics. Craik, a brilliant psychologist, is remembered as an influential genius (rather similar to Turing) who tragically died at the age of 31. In his short book (Craik, 1943; see Sherwood, 1966), he introduced the idea of modeling mental processes in the brain with mechanisms from science and engineering.

Cybernetics is a multidisciplinary topic that deals with information and control. It is particularly concerned with building models, particularly to explain adaptive behavior in animals and machines. The classic model is that of feedback, as seen when a ship's compass controls the rudder by being linked to it through the steering machinery. Even simple feedback mechanisms such as this are quite impressive in operation; the ship automatically adjusts to all manner of disturbances and constantly strives to bring the vessel back on the set course. These feedback loops are found in many devices, and the types that constantly track some signal, such as autopilots (rather than switches, like thermostats), are especially appealing to our sense of perceived autonomy and goal-following actions.

The early cyberneticians were inspired by this autopilot function (these are often called *teleological mechanisms*, or "devices that have a purpose"), and they realized that similar mechanisms might account for various phenomena in the human body and brain. Mammals can control their vital parameters within very fine limits: body temperature, blood chemistry (acidity and glucose), water and salt levels, cell renewal rates, oxygen levels, and even sleep quotas; and all these parameters are managed by intricate feedback mechanisms. The brain also has many adaptive, feedback-like characteristics, and the idea of modeling them with proven mechanisms from machines was, and still is, very attractive.

Cybernetics aims both to build theories of human behavior and to produce working models or simulations that display the same behaviors. This may sound like AI; the difference is that *all* the behavior should match human processes, not just the end result. This means that the mechanisms used in cybernetics must model the known biological data, or at least conform with current putative explanations when there are no data. As Donald Mackay said, "How far is it possible to envisage a model that would imitate human behaviour, and also *work internally on the same principles* as the brain?" (Mackay, 1954, 261, italics in original).

The most significant event in the cybernetics movement in Britain was the Ratio Club (Husbands and Holland, 2008). This was an informal discussion group that met regularly from 1949 to 1955 to explore the exciting cybernetic concepts of the day. Membership of the club was limited to about 20, was by invitation only, and was intended only for young scientists; it excluded fusty professors on the grounds that rank would inhibit discussion. The club set the career and research agenda for most of the participants, many of whom went on to become very distinguished scientists. The list of members includes such luminaries as W. Ross Ashby, Horace Barlow, I. J. (Jack) Good, Donald Mackay, William Rushton, Alan Turing, Albert Uttley, and W. Grey Walter.

Interestingly, the club knew of Norbert Wiener's work in the United States but did not follow it. The criterion for membership allowed only members "who had Wiener's ideas before Wiener's book appeared."[2] However, they did hold Kenneth Craik in great esteem. Unfortunately, Craik died in an accident before the club was founded, but his ideas were a major influence on the members. His reach went even further, as in the United States, Norbert Wiener and Warren McCulloch both recognized his contribution, and when AI was first established (as generally agreed) at the 1956 Dartmouth Summer Project, Craik's influence was very evident and explicitly acknowledged.[3]

The topics discussed by the Ratio Club ranged across biology, psychology, philosophy, and mathematics, but the focus was always on brains and machines and the fascinating relationship between them. Ross Ashby, a psychiatrist, built experimental homeostatic machines (which maintain complex equilibrium in the face of large disturbances) and wrote *Design for a Brain* (Ashby, 1976); Horace Barlow was a neuroscientist who discovered visual neuron codes; Donald MacKay, a physicist, wrote the book *Information, Mechanism and Meaning* and the article "Mind-like Behaviour in Artefacts" (Mackay, 1969, 1951); and Grey Walter wrote the book *The Living Brain* and the article "The Brain as a Machine" (Walter, 1953, 1957).

Grey Walter was a physiologist who also created the first autonomous mobile robot. His "tortoises," known as Elsie and Elmer, were motorized, wheeled platforms controlled by thermionic valve electronics and light sensors.[4] They could steer around obstacles, search for light sources, and return to a kennel for recharging. They were widely publicized in the media and exhibited as future technology at the Festival of Britain in the summer of 1951. All this took place by the end of 1951—a remarkable achievement! As Holland says, "Grey Walter emerges as a fascinating and far-seeing figure whose work has not dated, perhaps because it was explicitly based on similar principles to those used today in biologically inspired robotics" (2003, 2085).[5]

The 1950s was a decade of much fascination with "independent automata." Computing technology was being developed rapidly, and public awareness of the possibilities of modern machines was kindled by media excitement. In addition to Grey Walter's tortoises, many other synthetic animals were created, including robot squirrels that foraged for artificial nuts, robot mice that would run mazes, machines that played games, machines that demonstrated Pavlovian learning and associative memory, and machines that performed logical reasoning and inference.[6]

A fascinating example of such a synthetic animal is the robot dog built as a demonstrator by the giant Philips electronics company in the Netherlands in the mid-1950s (Bruinsma, 1960). The size of a large dog, it contained four big racks of electronics (using around 100 thermionic valves), banks of batteries, and lots of sensors. It had photocells for eyes with stereoscopic tracking circuitry, stereophonic hearing by two ears that could turn the head toward sounds, tactile switches with relative temperature sensing so that the robot dog could avoid objects and sense humans by their warmth, and an acoustic radar system for navigation. The dog could be called by a 400-Hz whistle, and it would avoid obstacles, stop when its name was called, and turn its head toward warm "food" and activate its tongue. It also could walk backward and wag its tail, and the acoustic radar system "produces a barking noise" (Bruinsma, 1960, 53).

Cybernetics quickly became an international and collaborative movement. Pierre de Latil's book *Thinking by Machine,* translated from the 1953 original *La Penée Artificielle* (de Latil, 1956), is a French classic covering feedback systems, theoretical models, synthetic animals, and other international work of the time. As the field became established, international conferences were frequently held and global interactions and collaborations became common. For example, the Josiah Macy, Jr. Foundation supported a regular conference series, chaired by Warren McCulloch, from 1946 to 1953. A number of cybernetic societies still exist today.[7]

MODERN CYBERNETICS

Nowadays, the word *cybernetics* has lost its original meaning. Despite being at the very heart of AI in the beginning, it gradually faded as a field of study and was overshadowed by AI. This came about partly because the topic eventually broadened and extended into areas such as economic and political systems, population models, and the social sciences generally. Topics such as "systems thinking" emerged and extended into management and other soft-science applications. At the same time, the treatment of key mathematical concepts, such as feedback theory, communication theory, and artificial neurons, gradually migrated into the harder sciences, such as control theory, engineering, and also psychology.

Some cybernetic outposts existed in university departments. In Britain, these included Keele University, the University of Reading, Brunel University, King's College London, the University of Bradford, and Bangor University. At Keele University, Donald Mackay founded the Department of Communication and Neuroscience in 1960. At the University of Reading in 1964, Professor P. B. Fellgett became the first professor of cybernetics in Britain. The department of cybernetics (and, later, engineering) at Reading taught cybernetic courses, several of which are still offered today, but a degree in cybernetics is no longer offered. It was a similar story across other universities worldwide. Although France and some other European countries still fly the flag, the word *cybernetics* has now been overtaken by terms with purely digital meanings (e.g., *cyberspace*) and has lost its original connection to biology.

Perhaps the most famous proponent of cybernetic ideas in recent years was one of the founding fathers of AI; Marvin Minsky. In 1970, Minsky co-founded and was the first director of the world-famous MIT AI Lab. Minsky, who died in 2016, might have denied any connection to cybernetics (he hardly ever mentioned the word in his writings), but he was often critical of the uses of formal logic and language in AI, and he had a biologically sympathetic outlook toward machines. Minsky didn't agree with purely symbolic systems, and his many interventions can be seen as steering AI toward models of biological and psychological phenomena that took a more holistic view of the problem. Informally, AI split into two camps: the "Neats," led by John McCarthy, who promoted the crispness and elegance of logical formalisms; and the "Scruffies," led by Minsky, who were willing to recognize the messiness of biology. See Minsky (1991) to learn more about Minsky's view on this topic.

A notable modern cybernetic development came from two Chileans, Humberto Maturana and Francisco Varela, who stressed the importance of living systems as the substrate for feedback mechanisms, and ultimately intelligence (Maturana and Varela, 1991). They created the term *autopoiesis* to refer to systems that can maintain themselves through cell renewal and other biochemical processes. This work is still inspirational for many, but it has been difficult to reconcile with, or translate into, conventional digital technology.

SYMBOL GROUNDING

AI has been firmly based on the symbol-processing model for most of its history. Right from the start of AI, the physical symbol system hypothesis was strongly promoted. This states essentially that human problem-solving and intelligence can be understood as the processing of high-level symbols. For many years, this credo was followed, assuming that logical symbol manipulation was sufficient to produce intelligence. Thus, only computer power and lots of programming were required, and hardware and robots could be ignored. Cognitive science also adopted this stance and has long treated cognition as the manipulation of internal symbolic representations of external events. Even artificial neural networks (ANNs) have not affected this view very much, but they have developed, in parallel, as a subsymbolic approach, along with other so-called soft computing technologies (like fuzzy reasoning, genetic algorithms, and other approximate computing methods).

Challenges to this tradition came in the form of the symbol grounding problem. Questions are often raised about the meanings of words and symbols. How can a collection of symbols in the head relate to real-life events in the physical world? How do symbols get linked to their meaning? If you ask someone the meaning of a word, the reply will be another set of words (this is the function of a dictionary). But imagine that you wish to learn the meaning of a Chinese symbol and you only have a Chinese dictionary. Then the definition of the symbol will be given in other Chinese symbols; you might try looking up those symbols too, but this would begin an endless trawl through more and more new symbols, never finding any meanings. This is the symbol grounding problem, as introduced by the cognitive scientist Stevan Harnad in a seminal short paper (Harnad, 1990). The problem applies to both brains and computers and directly challenges the assumption that symbol processing in a computer can produce intelligence.

The solution proposed for the brain is that some symbols are elementary symbols that have meanings in terms of sensory-motor perceptions and experience. That is, elementary symbols must be grounded in concrete experience gained through interactions in the environment. Higher-level symbols can then be defined in terms of the elementary symbols. In this way, all symbols can be grounded, or given meaning, directly or indirectly, from experience.

Even linguists have reinforced this view, stating that *all* words and concepts, including the most abstract, must have their meanings ultimately grounded in sensory-motor experience (Lakoff, 1987). For example, in the statement, "we are very close," an abstract relationship, friendship, is being described using a concrete word from physical experience, which itself is learned from direct experience of the proximity of objects in the environment by vision or other perceptual means. Here is another example: in the statement "grasp this idea, try to get a good hold of it," an extremely abstract concept (an idea), is being discussed in terms of one of our most concrete sensorimotor experiences.

This is a very significant change for the philosophy of AI. From the beginning, it was assumed that abstract thought was the key to intelligence, and so intellectual games like chess could provide all that was necessary to demonstrate and test AI. It followed that the manipulation and processing of symbols could represent events and objects in the external world, and cognition is rational and can follow logical reasoning patterns. (We know now that humans are far from rational, and symbolic and rational thinking, such as math, can be produced from nonrational brains.) All of this supports the physical symbol system hypothesis, provides the framework for nearly all symbolic AI, and backs up the notion that the brain is (just) a machine, so the mind can be simulated or reproduced on a computer.

THE NEW ROBOTICS

The concept of grounding has had profound implications for robotics. The early AI view of robotics was far too simplistic, consisting of a three-stage loop: sense–think–act (derived from the old standard computing model: input–compute–output). The "thinking" part gained all the attention, and so symbolic AI methods dominated, with mathematical logic, reasoning systems, theorem-proving, and other techniques being tested on robots. Eventually, it was realized that sensing and acting were intrinsically involved in thinking and could not be parceled out into separate modules.

It then became important for AI systems to be situated, embedded, and embodied. *Situated* means fully interacting with an environment, *embedded* means being a part of the whole system, not remote from the environment; and *embodied* means having a body. However the concept of embodiment is much more significant that just having a body. Embodiment means that the brain is not independent of the body and many aspects of perception, thinking, and acting are influenced by the nature of the body in which that brain resides. This means that the actual structure of the robot's (or person's) body and environment, and the robot's (or person's) interactions with the entities in that environment, have a profound influence on the thinking processes that are possible and on the developmental trajectory of intelligence.

A robotics pioneer in this approach is Rolf Pfeifer, whose view is clear from the title of one of his books: *How the Body Shapes the Way We Think: A New View of Intelligence* (Pfeifer and Bongard, 2006). Rodney Brooks helped to turn the tide by building some insectlike devices to demonstrate that adaptive behavior need not have high levels of computation, or indeed any symbolic processing at all, and argued impressively that thinking agents must interact with the world through a body (Brooks, 1991). Other robotics researchers followed this line (such as Randall Beer), and we now have a much more balanced and holistic view of the problems of robotics.

The embodiment idea is known in psychology and philosophy as *embodied cognition*. For readable discussions of its wider significance see philosophers like Andy Clark (1998), while neuroscientists like Antonio Damasio have identified brain regions and mechanisms that show how the body is intimately connected with the mind (Damasio, 2000). Damasio also gives an excellent account of how consciousness might emerge in the brain, but we will avoid that topic for now.

It can be a surprise to realize that the actual shape and structure of our bodies—the muscles, limbs, eyes, skin, and senses—all influence and mold our mental development and our thought processes. It's similarly remarkable that a creature with a different body will have different thoughts, simply because its actions and interactions will entail different experiences, and therefore different cognition. Of course, one of our primary experiences is interaction with other humans, and this allows us to build understanding and mental models of others, which is also a major part of our intelligence.

The more general term, *enaction*, was introduced by the psychologist Jerome Bruner, who saw action as one of three ways of learning and representing knowledge (the others being image-based and symbolic) (Bruner, 1966). David Vernon (2010) offers a good introduction to enaction (for instance, see the quote at the start of this chapter). He goes on to describe its five key elements: autonomy, embodiment, emergence, experience, and sense-making. *Autonomy* refers to self-control; embodiment means being influenced by one's physical body; *emergence* is behavior that occurs through the dynamics of internal and external interactions; *experience* is the state of a system as a result of the history of its interactions; and *sense-making* is the ability to learn regularities or contingencies present in the system's interactions. By making sense of its experience, such a system tries to capture some regularities or laws in its interactions, creating the potential for prediction. The enactive approach concerns active learning through anticipative skill construction rather than (more passive) knowledge acquisition processes, and claims that shaping and guiding action are fundamental for all intelligent systems.

OBSERVATIONS

- Early cybernetics showed the way, producing some remarkable robots. These pioneers took a holistic view and focused on general animal behavior. When AI began, in 1956, it started narrowing from the adaptive, behavioral view to a more cognitive, brain-based approach.
- Robots engineered for tasks will not be humanlike. The key to humanlike robots is growth in behavior, not task expansion.
- All natural intelligence is grounded in experience, and all thought and language is built up through environmental interactions.
- Physical anatomy influences thought—the body has effects on cognition. Embodiment shapes the structure and meaning of the robot's (or human's) dynamic interaction with the environment, and the internal

structures somehow capture the totality of the experience gleaned over the developing agent's lifespan.
- A developmental approach should not adopt fixed formalisms (such as symbolic or neural methods); rather, it should concentrate on the more general mechanisms, following Craik, Minsky, and others.
- Enaction is a new paradigm in robotics that promises to break through the difficulties of the task-based, input-output approaches. Two important corollaries are that environmental interaction (with objects and with others) is an essential aspect of learning, and that experience builds through development. Therefore, ontogenesis (the development of an entity from the earliest stage to maturity) is crucial.

12

THE DEVELOPMENTAL APPROACH—GROW YOUR OWN ROBOT

Babies develop incrementally, and they are not smart at the start. We propose that their initial prematurity and the particular path they take to development are crucial to their eventual outpacing of the world's smartest AI programs.
—Smith and Gasser, 2005, 13

Since the turn of the century, a new and active research community has come into existence. Known as *developmental robotics* or *epigenetic robotics*, this relatively small but growing group of international researchers recognizes that ontogeny is an essential aspect of human cognitive ability and brings together many disciplines, including computer science and developmental psychology, to explore models of cognitive growth. Taking inspiration from infant developmental research, both theoretical and experimental, the aim is twofold: to discover new learning algorithms for robots and to produce putative mechanisms for infant cognitive growth.

Developmental robotics acknowledges the remarkable growth of sensorimotor skill and cognitive ability in early child development. Human infants are born completely helpless, but in the space of around a year, they master control over a complex musculoskeletal system, learn about objects as they encounter them, and engage in complicated social interactions with their caregivers. Rapid learning and cognitive growth continue through the early years. This is a display of intelligence growing in a most general way, and the aim of developmental robotics is to find principles and mechanisms behind such growth that might be used in robotics.

Developmental robotics explores all aspects of infant behavioral development, with a particular focus on the growth of perceptual, cognitive, and interactive competencies. The main areas of study are sensorimotor skills, including body models, action spaces, and manipulation skills; visual development, including spatial perception, object understanding, and action affordances; social interactions, including imitation, cooperation, and mutual behavior; and language, including speech, symbol grounding, and linguistic structures. For on overview on this field, several short introductory papers are available (e.g., Lungarella, Metta, Pfeifer, and Sandini, 2003; Prince, Helder, and Hollich, 2005), and for more detail on all the main topics, see the excellent book *Developmental Robotics: From Babies to Robots* (Cangelosi and Schlesinger, 2015).

Developmental robotics is now well established, with an integrated community of researchers, an annual international conference series, and a specialist journal.[1] The newsletter of the IEEE Cognitive and Developmental Systems Technical Committee is a useful medium for sharing information about the community.[2]

This chapter will outline the key ideas behind my research in this area, which are all based on the belief that trying to build models of child development and implement some of these putative mechanisms in robots is a productive way of approaching social robots and understanding intelligence. The next chapter describes how we implemented these ideas on a real humanoid robot, together with various experiments undertaken and the results produced. (Where I use the first-person plural, this is not the royal *we*; it refers to the local research team that did all the hard work!)

Given that humans are the prime examples of intelligence, there are two main sources of relevant scientific information for budding developmental roboticists: developmental psychology and cognitive neuroscience. There is a huge body of literature on child psychology, covering both theory and experimental data. Many aspects of behavior and cognitive abilities have been explored, and at all stages of human maturation. These studies give a rich store of data about how infants change their behavior as they grow. Psychological theories built on experimental results are not always in agreement. This is because psychology deals with the *behavior* of individuals (i.e., the external manifestation of internal states). Without full access to the internal factors that are causing the behavior there will always be room for different interpretations of observed activity. However, psychology has made progress in pursuing and identifying some very plausible theories.

On the other hand, neuroscience is also important because it deals directly with the structure of the brain and the internal states that drive behavior. There is an equally huge amount of literature on brain science, including experimental studies conducted on living brains. Of particular interest is the field of computational neuroscience, which studies brain systems by means of computer models that can be implemented to explore possible mechanisms and processes.

Neuroscience is more advanced in computer modeling than psychology, partly because neural brain structures can be more easily mapped into artificial neural models. In psychology, despite the considerable understanding obtained through experimental studies, the mechanisms that drive behavior are less accessible and often not yet defined well enough to allow testing by computer implementation. This significant gulf between psychological knowledge and computational models represents a massive opportunity for computer modeling and great potential for new robotic techniques. Also, while robotics can learn much from infant psychology, there is also some value for psychology in gaining insights from computer models.

Because working on brains at the neural level involves invasive techniques, most of the results obtained refer to particular individuals, on particular days. The experiments are usually performed on monkeys or rats, occasionally on adult humans (during brain surgery), and *extremely* rarely on infants. Consequently, this kind of research does not produce much developmental data. In the past, questions about how brains grow and change could only be inferred from in vitro studies of brain samples taken at different ages. This is now changing, with rapid advances in medical imaging and scanning technology that offer in vivo structural and functional brain imaging. However, these methods cannot record individual neural activity; rather, they measure more general properties of regions of brain material. So the main difference between psychology and neuroscience is that the former mainly deals with behavior, while the latter mainly deals with brain structure. Recent work is bringing these together much more, and developmental robotics is playing a part in this development. For an introduction to the integration of developmental aspects into cognitive neuroscience, see Johnson and de Haan (2015).

THE ROLE OF ONTOGENY: GROWING ROBOTS

The embodiment movement in robotics research emphasizes the key role of sensorimotor experience as a foundation, a substrate, for higher levels of learning and cognitive growth. The ideas of enaction and grounding entail much more than a simple linkage between sensorimotor events and symbolic representations of action and sensation—much more than just the capture of sensorimotor data in a form that can be processed symbolically. True grounding can only emerge during the process of constructing representations through sensorimotor interactions. This is essentially a *constructivist* view of development, in which the agent's own experience drives and shapes the internal understanding of the world that it inhabits.[3] Constructivism is a relatively modern approach in psychology that has become widely accepted, and, to an extent, it has mediated the "nature versus nurture" debate.[4]

We've come a long way, from designing traditional artificial intelligence (AI) software to proposing integrated brain, body, and environment systems. The keywords are *developmental growth, embodiment, enaction,* and *interaction*. We can summarize our agenda as a hypothesis:

Hypothesis 2 *Robots with humanlike cognitive behavior can't be engineered by designing software, but must be developed through constructive processes of enactive cognitive growth via interactions with the body, the environment, and other agents.*

Of course, *growth* here doesn't mean physical growth, involving size changes in the mechanical structures, nor biological growth, which is far too difficult and well outside our comfort zone. Physical growth, of muscles and neural systems, is crucial for living systems, but we can only approximate the biological reality with the current, limited robotic components and hardware. Nevertheless, the growth intended here, in the cognitive abilities of the agent, can be modeled to some extent. If we can understand just some of the adaptation and learning that allows immature infants to transform into accomplished children, then this will be extremely valuable.

Alan Turing actually considered this issue in his writings on AI in 1950. Writing about how to create thinking machines, Turing suggested the possibility of "growing" a solution as an alternative to programming or "building" a solution (Turing, 1950, 456):

> Instead of trying to produce a programme to simulate the adult mind, why not rather try to produce one which simulates the child's? If this were then subjected to an appropriate course of education one would obtain the adult brain. …We have thus divided our problem into two parts. The child programme and the education process. These two remain very closely connected.

Turing did not give a starting age for the child program (he said, "Opinions may vary as to the complexity which is suitable in the child machine" [457]) and the "course of education" can be taken in its broadest sense to include experience in general, but it is clear that he is talking about cognitive development: "In the process of trying to imitate an adult human mind we are bound to think a good deal about the process which has brought it to the state that it is in"(Turing, 1950, 455).

It's encouraging that people like Alan Turing, Kenneth Craik, and the early cyberneticians thought this approach was worth a try, even if it lost impetus when modern computer hardware and software technologies grew to dominate the research ethos.

SEQUENCES, STAGES, AND TIMELINES

If you accept the constructivist and enactivist views, then you must start at the beginning. As Vernon argues, one cannot short-circuit ontogenetic development; everything builds upon previous experiences (Vernon, 2010). If you take an intermediate starting point, then, to justify your scientific model, you have to identify what competencies and knowledge the system already had available. This gets more and more complicated and difficult as the starting point becomes further and further advanced. Thus, in order to minimize the assumptions we have to make, we should start at an early a stage as possible.[5]

In line with this "start at the beginning" argument, our work begins with newborns and follows their sensorimotor development over the first year. Everyone knows how rapidly newborn babies seem to change from vulnerable feeders and sleepers into highly active and creative infants and toddlers. Actually, although it may seem fast, the couple of years needed for human postnatal development is an extraordinary long period compared to other altricial animals (i.e., those that need care from birth). The nearest animal equivalent (the great ape) reaches the same stage of development as a two-year-old child in only about a quarter of the time. This notable human characteristic of stretching out early maturation must surely be related to our eventual enhanced cognitive abilities.

During life, we progress through various stages, but in infancy, these are very pronounced. All children display a sequence of development wherein some competencies precede others. An infant will be seen to struggle with a task for a long while and then suddenly achieve some new level of skill and move on to a new pattern of behavior. This has been described as periods of growth and consolidation, followed by transitions toward the emergence of new behavioral phases.[6]

Of course, individuals vary considerably in the onset and duration of these sequences; some talk before walking and vice versa, but a trajectory of staged growth is always evident and appears to be very significant for the development of human cognition.

Historically, the pioneer in theories of staged growth was Jean Piaget, the famous Swiss clinical psychologist. Not all of Piaget's ideas have survived, but he is widely remembered for his foundational work on child development. His emphasis on the importance of sensorimotor interaction, staged competence learning, and a constructivist approach is particularly relevant (Flavell, 1963). Piaget called the first two years after birth the Sensorimotor Period, which consists of six subperiods that cover concepts such as motor effects, object permanence, causality, imitation, and play. These topics clearly have much relevance and interest to robotics.

Most research in developmental robotics deals either with specific behaviors, such as imitation, active vision, perceptual and motor skills, self-awareness, and social interactions; or with fundamental issues, such as knowledge grounding, the development of representations, motivation, modeling, and philosophical concerns. These studies often relate to cross-sectional data, in which a particular skill or behavior is examined across a group of subjects. There is less research on longitudinal development, looking at the cumulative effects of continuous growth and the totality of the developmental trajectory. Our research interests lie in exploring mechanisms that could correspond with early infant cognitive growth, with a focus on the general, gradual acquisition of sensorimotor skills. So we believe that longitudinal studies, covering the growth of various bodily competencies over significant time periods, are very important.

Longitudinal experiments require an understanding of normal developmental sequences and their temporal appearance. Fortunately, psychologists and clinicians have extensively studied maturing infants and produced various timelines of development. In addition, the research on infant development provides plenty of data on the fine details of individual behaviors and their structure. We carried out an extensive review of this literature and summarized the key findings suitable for robot implementation (Law, Lee, Hülse, and Tomassetti, 2011). This paper gives a breakdown of weekly development in the fetus and then monthly development in the infant, from birth up to one year. A series of charts showing timelines for sensory competencies and motor abilities were produced from this data. These charts illustrate the relative ordering of emerging competencies and their various interrelationships. Table 12.1 shows a simplified example timeline for motor control.

Another way of recording development is to follow a particular behavior and note all the prior competencies that have to occur to support the emergence of the behavior. Table 12.2 presents a chart for the ability to reach and grasp an object: a skill that is first achieved at around 5 months. This shows the various components in the sensorimotor systems that must mature before successful grasping can be achieved.[7] Notice the long duration of "hand regard" behavior; babies spend a great deal of time just staring at their hands! We found this an almost essential process in our robot experiments for correlating the very different spaces and calibrations of eye and hand; if you know that you are looking at your hand, then both systems are addressing the same location in space.

Table 12.1 A simplified timeline of the growth of motor control competence in the first nine postnatal months.

Motor system		Birth	1	2	3	4	5	6	7	8	9
Eyes	Pan, tilt	Control ――▷									
	Vergence	Vergence ――――▷									
	Eyelids	Working									
Neck	Roll, pitch, yaw	Control ――▷									
	Torque	Torque ――▷									
Shoulder	Roll, pitch, yaw			C	――	▷					
	Torque			T	▷						
Elbow	Pitch			C	――	▷					
	Torque					T	――	▷			
Wrist	Roll, pitch, yaw							C	――	▷	
Hand	Thumb opposed					r	――	――	▷		
	Thumb					r	――	――	――	▷	
	Fingers			J		I					
	Grasps					U		P	R	N	
Torso	Roll						C				
	Pitch				p	――	▷				
	Yaw					p	――	――	▷		
	Torque				T ▷						

Key: C, increasing control (from onset to effective function); T, increasing torque; r, increasing range or refinement; J, parallel finger movement; I, independent finger movement; U ulnar grasp; P palmer grasp; R, radial grasp; N, pincer grasp; p, increasing precision. Modified figure 2 from Law, Lee, Hülse, and Tomassetti (2011, 346), with permission from Sage Publishing.

These timelines specify the developmental progression seen in human infants, which can then be used as a plausible trajectory of behavioral competence that might be expected from a successful robot model. Other examples of timelines and road maps for robotics research include those from the iTalk project (Cangelosi et al., 2010) and the RobotCub project (Vernon et al., 2010).

CONSTRAINTS ON DEVELOPMENT

It is clear that some behaviors are impossible if the underlying systems are so immature that they cannot be controlled effectively. Looking at table 12.2, we can see that the requirements for successful reaching toward and grasping an object include building several prior component

Table 12.2 Infant behavioral stages prior to reaching and grasping an object, resulting from the constraints of immaturity.

Weeks	Sensory behavior	Motor behavior
4–6	Periphery vision sensitive, poor fixations	Limbs active, many reflexes
6–8	Short-duration fixations, some looking at hand	Stationary hand
8–10	Better fixations, much hand regard, better gaze shifts, stimulus pursuit	Stationary hand, swipes at objects with fist closed
10–12	Sustained hand regard, glancing between hand and object	Body activity with stimulus, swiping at objects
12–14	Some hand regard, glancing–hands and object	Bilateral arm activity, hands in contact on midline
14–16	Less hand regard, glancing at objects	Bilateral response to objects, some torso alignment
16–18	Glancing with slow approach	Fumbled grasp, hand opens for object
18–20	Object-fixated reach	Rapid arm reach, hand opens during approach

Note: This is a compilation of basic data from many published sources.

competencies: peripheral detection of the object, eye movement to fixate on the object, correlating the fixation point in space with the spatial target point for the hand, opening the hand, reaching out to the target point, and closing the hand upon tactile contact. If any of these competencies are undeveloped or immature, then the action may fail.

Timelines are a way of mapping out the dependencies between the various stages; the subsystem immaturities can be interpreted in terms of constraints on abilities. In developmental psychology, there are many theories on the origins, role, and effects of such constraints on development (Campbell and Bickhard, 1992).

However, although constraints may *appear* as restrictions on growth and learning, they can have important functions in actually facilitating development. Any constraint on sensing, action, or cognition effectively reduces the complexity of the inputs and/or possible actions, thus reducing the task space and providing a smaller framework within which adaptation or learning may progress.

When a high level of competence at some task has been reached, then a new level of task or difficulty may be exposed by the lifting of a constraint. A new phase then starts under the properties of the newly scoped task and eventually gains further competence by building on the accumulated experience of the levels before it. The psychologists Bruner (1990) and Rutkowska (1994) have produced some inspirational insights from their research on this topic.

It has been suggested that the unequal developmental rates produced by constraints may allow independent development of the subsystems, thus reducing competition among them (Turkewitz and Kenny, 1982). This may help the growing neural organization of perceptual and cognitive functions, and later, when individually developed, the subsystems can be combined and integrated to support complex behaviors.

Constraints on development originate from many sources and can be of several types. Here is a brief breakdown:

Anatomical/hardware These are the physical limitations imposed by the morphology of an embodied system. Usually these are experienced as restrictions on freedom of movement. Thus, the body structure will have limits on the possible angles of the skeletal joints, which in turn will give kinematic restrictions (think of scratching your own back). The skin, muscles, and organs can also act as mechanical obstructions, producing inaccessible regions of local space.

Sensorimotor All sensors have their own limitations; usually these are specified in terms of accuracy, resolution, response rates, and bandwidth. These parameters can change over time, and for babies, they are initially quite limited by their immature neurological and physiological structures, such as low visual acuity and restricted focusing of the newborn eye. Motor systems have similar characteristics, with additional limits on dynamic performance (velocities and accelerations) due to immature muscle tone.

Cognitive/computational Constraints on cognition take many forms, not just concerning speed, but also relating to information content and structure. A well-known adult example is the (small) size of short-term memory (try remembering a 10-digit phone number after a short interruption). Sensory constraints can also affect or restrict important input to cognitive processes.

Maturational All the constraints mentioned here are maturational to some degree, but this group refers specifically to those internal biological growth processes that are difficult to enumerate but still influence the general performance of both cognitive and bodily systems. Examples include endocrine influences, nerve growth and myelination rates, and neural support systems (e.g., glial cells).

External/environmental External constraints are those that restrict sensory input or behavior in some way but originate from the environment, not from the agent. These can be applied at any time and are not related to the individual's stage of growth. Positive examples are seen in parental care, education, and many other social situations, where the environment has been carefully arranged so that interactions are more likely to direct attention or action toward a desirable goal or outcome. This is often known as *scaffolding*.[8]

It is generally agreed that constraints or biases play important roles in learning, knowledge acquisition, and cognition generally, but how this happens is still very open (Keil, 1990).[9]

START SMALL AND START EARLY

There are two observations about maturation in human development that seem to form general principles. First, the motor systems of the newborn mature in a particular progression through the body—proximodistal, from the center out to the appendages, and cephalocaudal, from head to tail. Second, the sensory systems gradually mature from coarse to fine resolution (Turkewitz and Kenny, 1982). The newborn can only move its eyes at first, but this is followed by neck rotation, then arm reaching, torso control, and eventually progressing to walking. This progression of motor maturity follows a clear proximodistal pattern; even the arms start with shoulder movement, followed by forearm control and then hand and eventually finger skills. Sensory systems all seem to begin with poor accuracy, limited resolution, and restricted range, and then they gradually extend to reach adult levels by around three years of age. For example, consider vision; the newborn's visual acuity is about 40 times worse than adults, their visual field of view is narrow (about 30 degrees, compared with around 90 degrees for adults), and they have serious deficiencies in color vision and contrast detection. These structural constraints are governed by internal maturation rates, so their effects can be seen in the timeline trajectories.

It might seem extraordinary that when trying to learn about the world in all its complexity of colors, shapes, and patterns, the baby is provided with very blurry, tunnel vision and ineffective muscles! But the principle here is one of "starting small" and growing the understanding of the world, in concert with the increasing information available from the maturing sensorimotor systems. When we consider what the baby actually experiences, we see that these initial constraints on focus and visual range are tuned by evolution to just that region of space where the newborn has the maximum chance of seeing the mother's face during feeding. A crude form of face detection is fully functioning at birth, and once contact with the mother has been established, her image will be refined as maturation lifts the constraints and visual stimuli can be explored with more attention.

The coarse-to-fine resolution theme has been described as the importance of starting small, or, sometimes, "less is more." This has also been demonstrated in experiments using neural networks in language processing (Elman, 1993).[10]

It seems logical to start developmental experiments with the robot in a state equivalent to that of a newborn. But this is not a blank slate—we need to recognizes the sensory competencies and motor behaviors that have grown during the fetal stage. Some ignore these as reflexes that will soon be superseded, but they are important starting points for the bootstrapping that comes after. The initial conditions for complex systems are often crucial determinants of their trajectory into the future.

So what happens before birth? A total of 16 distinct motor patterns have been observed and described in the fetus (de Vries, Visser, and Prechtl, 1982). Apart from the many localized actions (like sucking, hiccups, and jaw movements), there are five interesting kinds of major movements: startle (fast and large), general (slower and smooth), head movements (rotation and flexion), isolated arm or leg movements, and hand-to-face contact. Vision is poor, with only the ability to detect diffuse, dim patches of light, while hearing is very well developed (e.g., voice recognition). The tactile senses are also well advanced, with a distribution of sensors over the body very similar to that of an adult.

In this limited situation, in the enclosed environment of the womb, the best opportunities for sensorimotor learning concern tactile-motor contingencies. As the limbs move, they will encounter resistance from the uterine wall and also produce self-contact events with the somatic sensors. In Tokyo, Yasuo Kuniyoshi and his colleagues have pursued a program of computer modeling and simulation, working on detailed computer models of the human fetus. Their simulation model consisted of a skeleton, muscles, spindles, tendon organs, spinal circuits, and a basic nervous system, all in a simulated intrauterine environment (Kuniyoshi and Sangawa, 2006). The resulting dynamics of the simulated neural/body system exhibited emergent motor patterns, producing meaningful motor behavior, all without any preprogrammed goal structures or coordinated control systems.

This research showed that nonuniform distributions of tactile sensors over the body produces more varied movement behaviors. In humans, the densest distributions of tactile sensors by far are on the face and hands, and this relates to emergent hand-to-face actions. They also showed that the jerky and isolated limb movements can emerge from general, diffuse behaviors. This work was consolidated and extended to include realistic (and rather scary!), physical, babylike robots (Yamada, Mori, and Kuniyoshi, 2010).

THE IMPORTANCE OF ANATOMY

The shape of our bodies, the length of our arms, and the quality of our eyesight all influence the personal, egocentric view of the world that we build up and maintain. This is central to our understanding of the world and, indeed, to our very thinking processes. The enactive approach says that the body, environment, and brain are all intimately and mutually involved in building up internal representations solely through active experience with whatever interactions the bodily systems can provide and support. As the internal representations develop, so they embody the history of experience that the agent has lived through. "Since cognition is dependent on the richness of the system's action interface and since the system's understanding of its world is dependent on its history of interaction, a further consequence of enactive cognition is that, if the system is to develop an understanding of the world that is compatible with humans, the system requires a morphology that is compatible with that of a human" (Vernon, Metta, and Sandini, 2007, 122).

For this reason, it is important that we experiment on humanoid robots with a morphology that is close to the human body. The systems of the human body have particular structures; for example, the reach space of your arms is smaller than and differently shaped from the visual space of your eyes. This means that we have to coordinate and correlate these noncongruent spaces so that we can pick up objects with different arm configurations as required, but also that we don't reach for items that we see but are beyond our reach.

Of course, robots don't have to be humanoid; they can take many forms. But designers need to think hard about what their robots will actually experience. What does it feel like to move around with wheels instead of legs and feet? What understanding of the world can you create if you are a mobile floor cleaner only 20 cm high? (Put yourself in the "shoes" of such a robot: Imagine looking out from a floor-level camera and having bump switches on your body. What would your world look like? What could you make of the experiences that you can perceive from your limited sensory and motor capabilities?) Certainly such morphologies will produce a very different model of the world, and it may be difficult for us to understand them.

Even in hand-eye systems, where a camera views a robot arm's workspace, if the camera is mounted at a distant point from the arm (as happens often in industrial settings), then the correlation will bring quite different meanings than the usual human perspective. Cameras mounted near the proximal end of the arm (i.e., the shoulder) will give much more compatible worldviews; pointing at objects, gestures, and other arm-based signals will have much more apparent meaning to an observer.

Hence, we must focus on humanoid robots because they will develop understandings that are more compatible with ours. We want to promote successful human-robot interactions; if the robot builds humanoid-based cognitive features, then we will have a much better chance of understanding and relating to them. This factor should also make the experiments easier to run and analyze, not least because there is plenty of data on humans that can help in evaluating the robot's progress.

THE AMAZING COMPLEXITY OF THE HUMAN BODY

When we effortlessly reach out to pick up a cup of coffee, we are entirely unaware of the control problems that the body and brain has to overcome. Each part of the arm has to be moved just the right amount and at the right speed to bring the hand into contact with the cup handle.

Most joints in the human skeleton have only 1 degree of freedom (DoF); that is, they can move in only one direction. For example, the elbow joint allows the forearm to extend away from the upper arm. Each DoF can be represented as a single variable (e.g., the angle at the elbow). There are many joints in the body (about 230) that allow movement, and they produce 244 DoFs (slightly more degrees than joints because some joints, such as the shoulder and the hip, allow several directions of movement). Things are even more complicated because the joints are moved by the muscles and there are several groups of muscles operating on each joint (totaling around 630 in all).

For the coffee cup, we only need to use one arm (7 DoFs) and a hand (27 DoFs). Even if we ignore the hand for now, 7 DoFs are more than needed. Space has three dimensions, so 3 DoFs are sufficient to define any point in front of the body. We see space as three-dimensional (3D): left-right, up-down, and near-far; and values for these three variables are all that is needed to define any location of the cup. But rather than 3D, our arms move in seven dimensions (7D)! They have redundant movement possibilities.

This has been a headache for control engineers for ages. What is known as a forward control model is relatively easy to use—for a particular position of the arm, if we know the angles for all seven joints, then we can apply some simple math to calculate exactly where the hand is in 3D space. Going from 7D to 3D is easy. But if we specify the 3D location of a desired hand position and want to calculate what the joint angles should be to get the hand there, then there are more unknowns (7) than necessary, and the math gets very difficult. This is known as the *inverse kinematics problem*. You can see this by trying various different ways that you can arrange your arm while still keeping your hand in the same location on the coffee cup. This demonstrates the redundancy in the human arm, which is also seen in all the other major components of the body.

This problem, known in modern data theory as *the curse of dimensionality*, was first investigated in the body by a Russian neuroscientist called Nikolai Bernstein. Bernstein wondered how the brain could control all these DoFs, in real time, while also coping with external forces such as gravity. He demonstrated the difficulties of the problem in a novel way: by adding extra DoFs to the body and asking subjects to control them. In his classic ski-stick experiment, a ski pole is attached to the waist of a subject, and the other end has two elastic cords tied on. The subject holds the end of one cord in each hand and faces a wall with a large square or circle drawn on it. The subject has to trace out the figure on the wall with the far end of the pole by using hand and waist movements. This is surprisingly difficult because the subject has to learn to control a (new) 4-DoF system, while the output is only two-dimensional (2D). Bernstein's book is still a classic (Bernstein, 1967).

But evolution finds ways of solving problems like this. Newborn babies don't try to control their DoFs all at once. If they did, it would become impossible to move all their muscles at the same time. Instead, they produce stereotypies—stereotypical patterns of movement: those waving, banging, and kicking activities that babies seem to enjoy so much (Thelen, 1979). These movements can cover large regions of space but make small demands on their developing control systems. Close inspection reveals that some muscle groups and joints are active, while others are not. Maturation of body systems occurs at different rates; although they are all maturing rapidly, they don't mature at exactly the same time.

Proximodistal and cephalocaudal sequencing progresses as maturation spreads down from the head and outward to the extremities. The eyes move first, and they come under control while the newborn can only lie supine. Then neck rotation moves the head, thus extending the range of the eyes and allowing some coordination with sound direction. After a few months, trunk rotation can be recruited to help in reaching, and later on, walking begins to open up new possibilities and extensions to the local space.

AUTONOMY AND MOTIVATION

We all want robots to be autonomous, if only for the simple reason that we don't want to be bothered with constantly telling them what to do. Naturally, we expect to give them some general instructions, such as "Make a pot of tea at five o'clock," but these instructions should be minimal and not include telling our robot *how* to do something.

Real-life unstructured environments, such as a person's home, are typically unreliable, inconsistent, and subject to change; unexpected events frequently occur. This is why autonomy is essential to deal with these issues. Clearly, very sophisticated adaptation and learning abilities are necessary to cope with novel experiences and handle them appropriately. Not only must current skills be adaptable to new contexts, but the robot must learn to create new competencies for entirely new situations. This requires a level of autonomy as yet unseen in robotics. Necessary requirements include cumulative, incremental, real-time learning, as well as some kind of internal motivation or drive.

Assume that we have managed to design a robot that can grow in competence by learning about its body and local environment: What will motivate it? When we first switch it on, what will it do? Why should it do anything at all? Following our approach, it is not task-based, and so it cannot have any built-in task or purpose (otherwise, it would stop when the task was completed). So we must avoid designing or programming in any way that might give a bias or direction toward any particular path. A traditional industrial robot would simply be unable to function in this way; it would do nothing: "no goal, no activity."

Reward-driven learning is very popular in AI, but this depends on the delivery of a reward when some goal or subgoal is achieved (e.g., reinforcement learning, as discussed in chapter 7). Such explicit signaling of goals solves the motivation problem, but this is extrinsic motivation and relies on external training by the designer. This raises the question: Who decides what is rewarding? We will eventually teach our robot through human-robot interactions, but in the beginning (equivalent to birth), the robot's system will not have the ability to interact effectively.

So our robot must have some intrinsic motivation that drives it to act when no clear indication of what to do is coming from the environment.

The use of random action or "motor babbling" (discussed shortly) has become popular in robotics as a way to jump-start a system and obtain some stimulus input. But through the enactive approach, we see that action amounts to much more than just "do something rather than nothing." Motor activity actually serves many functions: It produces immediate inputs from proximal and internal sensors (e.g., tactile, kinesthetic, proprioception), it provokes the external environment (thus gaining information), and it gives additional viewpoints on current situations (exposing variations and contingencies).

We use novelty as a source of intrinsic motivation. By *novelty*, we mean anything unexpected, something new, or any familiar thing in an unusual context or situation. So a new movement, a new sensory signal, an unexpected result of an old action, and a first use of any neural, sensory, or motor system are all considered novel. This is very closely related to curiosity and expectation, but my use of novelty can be best defined as the inverse of familiarity. As a developing system grows, the framework of novelty enlarges to match the complexity of its capabilities. We believe this relatively simple form of intrinsic motivation is sufficient to provide the driver for a robot to travel a long way along the developmental timeline.

Novelty- and curiosity-driven behavior is a very active research area in developmental robotics; for reviews, see Gottlieb, Oudeyer, Lopes, and Baranes, 2013; and Oudeyer, Baranes, and Kaplan, 2007. Intrinsic motivation is a major characteristic of human life: It exists in other animals but has less range. While avoiding philosophical discussions about free will, it is clear that novelty, play, and curiosity are important aspects of intrinsic motivation—and development itself.

PLAY—EXPLORATION AND DISCOVERY WITHOUT GOALS

We are all familiar with infants playing with toys, other objects, and their adult caregivers. Many young animals, especially mammals, exhibit play behavior, and in particular, human children get involved in quite extensive periods of play. But what is play? To most people, it's another obvious phenomenon, just like intelligence, but to psychologists, it can be a controversial topic. The literature ranges from in-depth scholarly studies to comments of "no useful function and not worth studying." So trying to review the various definitions of play is a bit of a nightmare because there are so many (at least 50), and their authors can be quite contentious.[11]

It seems the difficulty is mainly down to the diffuse forms that play can take. There is no unique play behavior that can be identified because play can occur across different activities and in different contexts. Indeed, it can occur in any context and cover all kinds of behavior. This means it is hard to produce definitive or concrete examples that everyone can agree on.

Nevertheless, it is clear that play does exist; Piaget describes play as an orientation or an attitude, not a behavior, wherein an infant repeats actions "for the fun of it" without expectations of results, and with obvious enjoyment.[12] (Piaget, 1999). In fact, play is a fundamental and *essential* component of early cognitive development in humans (Ginsburg, Committee on Communications, and the Committee on Psychosocial Aspects of Child and Family Health, 2007). From a distillation of the literature, we find that play is an intrinsically motivated, goal-free, motor-centric activity. That is, it is a voluntary, intrinsic activity (Smith, 2009) that is not focused on objects or concepts and is activity based. It is also obviously enjoyable (Bruner, Jolly, and Sylva, 1976) and often observed to be repetitive behavior, involving much audio, tactile, and visual input. This view also reconciles with the action perspective in psychology (von Hofsten, 2004), from which play is not seen as random or meaningless activity but rather as the purposeful seeking of enjoyable action possibilities. From all this, I suggest this definition of play:

Any enjoyable behavior pattern that occurs without apparent regard to external goals or context, is action centered, and is often irregular, with unnecessary repetition of some actions, exaggeration of others, and even the inclusion of irrelevant actions.

What does play mean for very young babies? Much of the activity produced by very young babies is called *motor babbling*, by analogy with the verbal babbling produced by prelinguistic infants, and is assumed to have a similar function. Motor babbling is seen in many behaviors: sucking, eye movements, head rolling, facial expressions, body and limb-kicking actions, reaching, and touching. This is often observed to be enjoyable and action-centered, and it seems to have all the hallmarks of play behavior. I see no evidence from the literature that contradicts this view, so we can say that the general phenomenon of play includes all forms of motor babbling. We can then hypothesize for robots:

Hypothesis 3 *Intrinsically motivated activity in the form of play behaviors (as seen in babies, infants, and older children) is a fundamental and necessary component for the development of truly autonomous intelligent agents, both human and artificial.* Play *is intended as a general term that covers a wide range of behaviors from early motor babbling, through solitary physical activity, to complex interactive and social scenarios.*

It is important to notice that, although play is a kind of exploratory activity, it is not the same as normal goal-directed exploration. In exploration (finding a way through a building or examining the parts of a machine), there is usually a goal or purpose, such as to find a particular room or learn why a device isn't working. Goals provide targets and focus attention on specific tasks. But play is more open ended and has a wider scope. Although it seems to have no goal or purpose, it is not meaningless or aimless; I think its overarching role is in goal-finding, that is, it discovers interesting possibilities that may become goals of further activities. Thus, play generates activity that may produce interesting events that can then become prospective goals for future generated activity. This answers another questions: Where do goals come from? Apart from external motivators related to survival needs, how can intrinsic motivation provide useful structured activity if there are no goals? What is a well-fed, well-rested baby going to do? By playing with objects and situations in new contexts, or applying tried and tested actions to different environmental events, new experiences and possibilities open up. Thus, desirable states are discovered through play, which then become established as *proto-goal*s and can be further explored in the form of "goal babbling" (Rolf and Steil, 2012). You can see why it is important to this approach that the designer should not set goals for the robot (at least, not built-in goals—any desired task can later be taught through interactions in the environment).

An interesting idea has emerged from heart-rate variability studies on infants (Hughes and Hutt, 1979) and brain-scanning studies on adults (Tang, Rothbart, and Posner, 2012). It seems there are a few gross

Table 12.3 Some gross brain states.

State	Behavior
Sleep, anesthetized	Unconscious
Baseline or default state	Resting, free thinking
Goal-directed activity	Problem-solving
Observational activity	Imitation learning
Goal-free activity	Play

patterns of brain activity that correlate with certain broad kinds of behavioral states. An unsurprising distinction can be found between the states of sleep and wakefulness, but there are also notable patterns that separate problem-solving (which includes exploration), imitation, and play behaviors (see table 12.3). Attention and relaxation levels are different in these three classes: problem-solving is strongly focused on a goal, often to the exclusion of other input; imitation pays great attention to the actions of another agent; and play occurs in very relaxed conditions, with strong internal motivation. Of course, it is early days for these speculations, but they do suggest that play may have its own particular function and offer an alternative (or complement) to the dominant goal/reward model seen in most AI.

AN ARCHITECTURE FOR GROWTH

We now need to bring all the topics in this chapter together to form the design for our growing robot. The key design issues that we must deal with include sequences on a timeline, relevant constraints, and a motivation algorithm.

First, we need to establish a baseline for the developmental sequences that our robot should go through. This is obtained by taking one or more timelines from the infant development literature and mapping them onto the robot's available sensorimotor hardware. Existing robot devices remain pale approximations of the biological ideal, and we can use only those features and functions that are available in current hardware technology. When the hardware has been selected or specified, the robot's motor systems and sensors are matched with the timelines to duplicate as much as possible. At the end of this process, we will have a robot timeline that gives desired behavioral sequences for the various available subsystems.

Constraints then come into play and are entered onto the robot timeline. These will indicate when sensory accuracy should be increased or when muscle tone can be improved. These events will not be recorded at specific time points; the timeline itself will not have an absolute temporal scale, but rather will represent relative relationships. The relative order and durations of the sequences are what matters; infant time points are not relevant in robot time. The constraints covered here are the structural kind, mainly the sensorimotor and maturational types. This includes proximodistal, cephalocaudal, and coarse-to-fine-resolution constraints and their relative schedules of improvement.

Finally, we need to design an algorithm that will cause the robot to act and to grow from experience in the manner described so far. The algorithm must be driven by intrinsic motivation (novelty being the only internal value signal), it must attempt to repeat and learn interesting actions and experiences, and it must generate behavior that builds new skills from experience and the use of its prior behavior base. It must also cope with the anatomy of the robot, including things like awkward configurations and redundancy (Bernstein's problem), but these are indirect issues. We must not build these into the algorithm; rather, they must be handled by the system almost as a by-product of the operation of the whole.

This may seem like a strange approach to anyone who has written any computer programs. But remember this is a *synthetic* experiment—we don't expect it to work perfectly the first time, but we will run it over and over again, each time adjusting the design or the constraint parameters to get closer to the desired behavior. Of course, all programs are run many times during implementation, but that is done usually to remove the bugs and check that the programs satisfy the specified requirements. In our case, it's not just the code we are debugging, but the whole developmental model.

Following many experiments, we can condense the essence of our developmental approach into an algorithm. (There is no benefit here in delving into the depths of various computer programs.) We can describe the important features of the system in terms of a set of rules that indicate

Table 12.4 The developmental algorithm: key behavioral components as rules.

	Condition	Activity
1	Notice familiar stimulus	Perform appropriate action
2	Notice novel event	Direct attention to event
3	Novel event is exciting	Act toward novel event
4	New experience gained	Attempt to repeat
5	Experience can be repeated	Store the details
6	Several events cooccur	Assume connected in some way
7	Novel events are rare	Begin play activity
8	Novel events are extremely rare	Lift a constraint if possible

how it should behave. Table 12.4 lists a series of situations showing how we might expect or desire the system to behave under various conditions. These "rules" are really more like guidelines, but they give the essentials of the basic algorithm. The first rule, *Notice familiar stimulus → Perform appropriate action*, simply says that a known condition/action experience can be performed when the situation arises. This is relevant only after some experienced relationships have been learned and stored. For example, pick up an attractive toy when seen, or shake a favorite rattle. Clearly, a memory is required to record such experiences, and we use an associative memory system, which will be described in chapter 13.

The rule, *Notice novel event → Direct attention to event*, requires some kind of novelty detector. We employ simple statistical counters attached to all sensorimotor experiences and their component parts. These are incremented on each appearance of a particular item; thus, large numbers represent familiarity and low numbers represent rarity. This gives a form of saliency to new items, and we use an excitation mechanism to attract attention and resolve multiple novelty signals (low familiarity gives high excitation). All excitation values decay with time, but they can be reenergized upon renewed interest. A winner-takes-all criterion selects the most excited item for paying attention to.

When a novel event has been noticed, or rather the *most* novel event has attracted attention, then the system is motivated to act in some way *toward* that event. This can be done by transferring the high excitation level of a stimulated subsystem to all other modalities and subsystems. For example, if a new stimulus is signaled in peripheral vision, then the eye may saccade (very rapid eye movements from one gaze point to another) to fixate on that point, and if the fovea (the part of the eye that provides the clearest vision) then senses a new experience, that might be novel enough to raise excitation to cause further action, such as reaching toward the stimulus. This is the rule: *Novel event is exciting → Act toward novel event*.

New experiences can be the result of coincidental circumstances or fleeting events that have no meaning. So a good test of reliability is simply to attempt to repeat the experience. This is the reason for rules 4 and 5. If the experience is repeatable, then it indicates some regularity and hopefully captures some of the "lawfulness" of the physics of the world. These experiential events are stored in the memory, becoming available as responses to future events (i.e., rule 1). All the memory contents are also logged with statistics, so that unused or unsuccessful memories become less likely to be used in the future. This means that any faulty experiences that get stored will eventually become defunct.

The human perception of items or events that are close, either in time or space, as being one and the same is captured in rule 6: *Several events cooccur → Assume connected in some way*. Humans perceive any events that occur within about 10 milliseconds of each other as being very tightly linked; for example, the flashing lights on road signs and police cars can appear to move from one place to another if they are synchronized so that one light goes out exactly as another one comes on. This process of things that occur together is known in neuroscience as "neurons that fire together get wired together."[13] This phenomenon seems to be a general heuristic of perception and occurs with spatial effects as well as temporal. The neuroscientist Horace Barlow calls these events "suspicious coincidences" (Barlow, 1985, 40): "elements that actually occur often, but would be expected to occur only infrequently by chance." We will see how these connections are detected in the next chapter.

The intrinsic driver in our system is to produce action. If nothing interesting is happening, then the system should perform an action. In other developmental projects, motor babbling is usually employed at this point, with more or less random actions being executed. In our case, we use the memory as a play generator. The idea in the rule *Novel events are rare → Begin play activity* is that action will produce something interesting. Our play generator is very simple—it extracts from memory an action of a previous experience that best matches the current context, while allowing some variation on the action. At the very early stages, the memory will have very little stored experience and the actions produced for play will be very primitive and may even appear as almost random actions. Thus, play naturally subsumes motor babbling. In table 12.4, each rule can be used whenever its conditions apply, so the play generator may just produce one action before another event distracts attention, or it might produce a sequence of actions, possibly leading to a new or novel situation.

Sometimes it seems as though nothing interesting ever happens (we've all been bored!). In the developing infant, maturity comes to the rescue here because new possibilities arise with new muscle strength and new sensory powers. Hence, the rule *Novel events are extremely rare → Lift a constraint if possible*. Of course, when all possible maturational constraints have been overcome, the environment is the sole source of stimulation; at that point, the robot will be "bored" until some external event is perceived as stimulating.

We called this algorithm LCAS (which means "Lift-Constraint, Act, Saturate"). The idea was that a structural constraint on the timeline would be lifted and the system would then be driven by new experiences in its somewhat enlarged sensorimotor arena. Many actions would occur during these experiences (some of them new), and novelty would be explored and tested (as in rules 2 to 5). Eventually, there would be very few novel events left to respond to—all the statistics would indicate a degree of familiarity, and there would be low excitation levels as a result; the system has reached (asymptotic) saturation. At that point, the saturation state triggers the lifting of another constraint (rule 8), and the process continues.

The idea of saturation has various implications. Excitation may be generated by unfamiliar stimuli from any of the subsystems (vision, tactile, proprioception, etc.) and so, for saturation, they must *all* be quiet. Thus, there should be a global measure of saturation, such as "Global excitation equals the sum of all local excitations." This global signal can then be the trigger for lifting constraints. In effect, this signal detects qualitative aspects of behavior, such as when growth changes have effectively ceased or when no new skills have been learned.

OBSERVATIONS

- Developmental robotics takes inspiration from the remarkable growth of sensorimotor skill and cognitive ability seen in child development. This is notable both for the speed of learning (often based on only one or two experiences) and the increasing difficulty of the problems overcome.
- Behavioral stages are fundamental in infant development and are consequences of maturation. As the various subsystems mature, constraints are lifted and new qualitative stages of behavior are observed.
- Timelines give guidance on behavioral sequencing and insights on the constraints of immaturity. They are valuable sources of information for robotic experiments.
- Humanoid anatomy is important for at least two reasons: It has an influence on the shape of our experiences and mental models; and similarity with humans engenders similar world understanding, and therefore better social empathy.
- Infant research suggests several guiding principles for robotics: Start at the beginning—you can't short-circuit development. Start small and simple—then grow in resolution and complexity. Start coarsely—then refine with experience. Start locally—build out from proximal to distal (including cephalic to caudal).
- Keep things as transparent and as simple as possible. Apparently simple algorithms can be surprisingly complex in their execution and consequences. As a corollary, beware of overcomplicated cognitive architecture. If we don't fully understand something at the beginning, things will only get worse.
- Our three hypotheses given in this chapter and the previous one argue that general, open-ended, goal-free intelligence cannot be designed as AI processes; rather, they should be approached through synthesis experiments using mechanisms such as intrinsic motivation and play generators. The key concepts are synthesis, constructivism, and enaction.

13

DEVELOPMENTAL GROWTH IN THE ICUB HUMANOID ROBOT

Play is the highest form of learning in early childhood. It is in play that children show their intelligence at the highest level of which they are capable. Play opens up new possibilities in thinking and develops the emotional intelligence that makes feelings manageable. It helps a sense of self and relationships with others to deepen.
—Tina Bruce, *Learning through Play* (2001, 112)

The previous chapter outlined, in broad terms, a new paradigm for creating intelligent robots. However, developmental robotics is not just an idea or a proposal; it is actively being pursued in many research labs. This chapter gives examples of concrete progress to show you that developing robots is not just a philosophical dream—such robots are actually being created and have a bright future. My own interest is in longitudinal experiments that produce learning over a range of competences. The work described in this chapter illustrates and supports this approach.

The details of robotics work tend to get very technical, and I don't wish to drown you in fine detail. I will only describe the results of our experiments in order to show what our robot achieved, and give just enough explanation of what was involved to give a flavor of how it works. Of course, I'm very eager to encourage people to dig deeper and get further involved, as there is so much promising work to do! For those who wish to know more, key references are cited for all the main topics, and an appendix at the end of this book summarizes what has been learned in terms of some general principles, and gives advice distilled from hard-won experience of working in this area.

These experiments show how a robot can start with absolutely no control, having only gross reflex actions equivalent to prenatal activity, and then, through a process of competence growth, reach a level of cognitive coordination that supports perceiving, acquiring, and playing with stimulating objects.

The main features of this approach are the role of maturational constraints on the learning and interplay of body and cognition; the significance of order and starting points in the developmental timeline; the incremental gain of control over multimodal, multidimensional systems; and the importance of play and intrinsic motivation.

ICUB—A HUMANOID ROBOT FOR RESEARCH

One of the principles of enaction is that the shape and function of the body affect the shape and function of the thinking that can emerge from that body. As David Vernon said, "The system requires a humanoid morphology if it is to construct an understanding of its environment that is compatible with that of human cognitive agents" (2010, 94). The iCub humanoid robot was designed with this in mind and is closely modeled on the anatomy of an infant of around 3 years of age [see figure 13.1 and Metta, Sandini, Vernon, Natale, and Nori (2008); Vernon, Metta, and Sandini (2007)]. iCub is the brainchild of Giorgio Metta and his colleagues and was created, through European Commission funding, at the Italian Institute of Technology in Genova. Around 30 robots have been built and are in use in research labs across Europe. The iCub has a total of 53 degrees of freedom (DoFs), all driven by electric motors, and all the joints have angle encoders that provide signals for proprioceptive sensing. There are charge-coupled device (CCD) color cameras in its two eyes, which can scan at realistic human speeds, and touch sensors are fitted in the hands, fingers, arms, and parts of the torso. Simple facial expressions can be produced by illuminating arrays of light-emitting diodes (LEDs) in the shape of a mouth and eyebrows. For details

13.1 The iCub robot grasping some soft objects.

about the iCub open-source design, the hardware components, and the software support environment, see Natale et al. (2013).

The iCub has three sensory modalities: vision, touch, and proprioception. *Proprioception* is the sensing of the positions of the joints in the body—the feeling of where the various parts of our skeletal structure are in space. When gravity or other external forces come into play, the muscles may generate opposing and corrective forces, but the sense of position provided by proprioceptive sensors always reports accurately on the spatial locations of body parts. The proprioceptive system of the body is one of the most vital of all our senses, and yet it is significantly underrated, probably because it operates more subconsciously than vision.[1]

The iCub has precision angle encoders fitted to all its joints, producing accurate feedback on the positions of the body parts. A series of tactile sensors in the skin of the iCub, particularly in the hands and fingers, sense when objects have been touched and are important during grasping and other hand actions. The other sensing modality is vision. The human vision system is quite different from digital camera imaging, and special attention is needed to simulate infant vision, as will be explained later in this chapter.

MANAGING THE CONSTRAINTS OF IMMATURITY

Following the infant timelines in the previous chapter (shown in figures 12.1 and 12.2), it is clear that eye movements are the first actions to come under a human being's control. Next comes movement of the head, followed by the arms, hands, and torso: This is the head-to-tail (cephalocaudal) pattern of maturation in the infant.

From the developmental timelines, it is possible to map these patterns onto a robot system. Most of the changes in motor performance are due to muscle maturation–either gradual increase in muscle tone or activation of a new muscle grouping. As we are not able to accurately model this type of change, we simulate it as a series of constraints preventing or restricting movement by various sets of joints. First, the relevant maturational constraints are identified for the robot's sensory and motor capabilities. Then, by deciding when the constraints should be applied or removed, we can produce a chart that relates to the timeline sequences. This type of analysis is done for a range of behaviors, each time noting the points where a component starts to become active and when it is fully functioning.

It is also important to define what functions are available at the experimental starting point, equivalent to "birth." Taking data from various sources (for instance, de Vries, Visser, and Prechtl, 1982; and Kuniyoshi and Sangawa, 2006), we can assume that all the tactile sensors are fully functional at birth, and the proprioceptive sensors in the joints are also well exercised and operational. In newborn babies, the vision system is undeveloped, but phasic signals, from flashing or fast-moving stimuli, are readily detected. In the robot, the eye pan and tilt muscles should be active but unable to perform vergence for depth perception (*vergence* is the simultaneous movements of the eyes required for stereoscopic vision), and the arms and legs can give stereotyped, synergistic movements (such as all joints moving together).

When all this detail has been considered and incorporated into the timelines, we obtain table 13.1, which shows the conditions for the release of the constraints relevant to our robot hardware. This is not a schedule; it is important that the constraints are lifted by the state of maturity of the system, rather than by any timetable. The time points in "months" are only intended to indicate the expected ordering.

Table 13.1 Constraint release table showing the conditions for when the various motor and sensory systems start to become fully active.

System	Component	Constraint release condition
SENSORS		
Proprioception	All parts	Birth
Tactile	All areas	Birth
Eyes	Periphery sensitivity	Birth
	Brightness	Birth
	Acuity	Gradual over 9 months
	Field of view	Gradual over 9 months
	Color	Gradual over 3 months
	Edges	Gradual over 3 months
	Vergence	4 months
MOTORS		
Eyes	Pan and tilt	Birth
Neck	Roll	When eye saccades mature
	Yaw	When eye saccades mature
	Pitch	Not used
Arms	Shoulder	Birth: synergistic reflex
	Elbow	2 months: locked elbow
		4 months: all joints independent
Hands	Closed	Birth
	Open	3 months
	Fingers	5 months
Torso	Roll	4 months
	Yaw	5 months
	Pitch	Not used

Note: The timescale refers to "birth" as the start of the experiment, and 'months' are simply relative periods of time. Modified figure 4 from Law, Lee, Hülse, and Tomassetti (2011), with permission from Sage Publishing.

VISION, GAZING, AND FIXATIONS

The human visual sensor, the retina in the back of the eye, is not covered uniformly in sensing cells, like the pixels in digital camera chips; rather, it has a high concentration in the central region, called the *fovea*. This is why the eye performs very rapid shifts to bring any stimulus of interest to the center for detailed examination. These ballistic movements, known as *saccades*, are driven by specialized neural circuitry that engages once the targets have been selected. So the eye acts more like an active searchlight than a stationary passive receiver, and this skill has to be learned by our robot.

When a stimulus is received on the retina, it will either land on the fovea, where it can be processed immediately, or (usually) on the periphery, which only contains larger, monochrome sensing receptors. The periphery of the retina is very sensitive to phasic stimuli (flashing or moving), and if something attracts our attention, then we need to move our eyes to bring it to the fovea for closer examination. We use a technique that learns the relationship between the location of a target detected anywhere in the periphery and the parameters of the eyeball motors necessary to drive the eyes to fixate on the target. Let's call this relationship R_{eye}.

Table 13.1 shows that brightness and periphery motion detection should be fully operational at the start of our experiments. However, the field of view and acuity of the eyes are subject to gradual improvement (coarse to fine) over the whole experimental period. This means that, although brightness is functional, all that can be seen at the beginning are fuzzy blobs of locally increased intensity. Figure 13.2 shows a series of images from the robot's eyes (i.e., cameras) that have been processed to simulate improving visual acuity. Color and edge detection have no function at the lower levels of granularity, but they gradually become influential as the field of view and acuity improve. These are examples of indirect constraints—we do not need a maturity condition for color and edge detection because they cannot contribute until other systems give them signals of a sufficient quality.

So when the robot starts, only the eyes are active, but it has no control over where to look, and the images sensed are very crude blobs of light and shade. Following the developmental algorithm described in chapter 12, in the absence of any stimulus, the robot should play using remembered actions related to the current situation; or in the absence of suitable memories, then motor babbling should be performed. For the eyes, *motor babbling* means random saccades with the aim of bringing the stimulus to the foveal region. Initially, there are a lot of misses, and several moves are involved before the target is found. But every miss adds information about the relation R_{eye}, and the number of saccade steps soon

13.2 The simulated maturing of visual input. Images widen and sharpen each month from birth to 9 months. The foveal region is indicated by a circle. Images from figure 2, Giagkos et al. (2017), with permission from the Institute of Electrical and Electronics Engineers (IEEE).

decreases. The criterion for releasing the neck constraints is when the average number of steps per saccade reduces to less than 2.

The directions of the eyes are not only determined by the eyeballs, but they are also affected by rotation of the head. Thus, both the head and eyes need to be coordinated. When the neck constraint is lifted, the effect of head movements on the eyes can be learned in another relationship, R_{head}. The local space that can be seen when using eyes and head together is considerably expanded, and we call this the *gaze space*.

MOTOR AND VISUAL SPACES

The motor systems of the arms, hands, and torso are brought into play at different points. Their constraints are lifted by implementing thresholds based on the progress of the other subsystems. Thus, arm activity

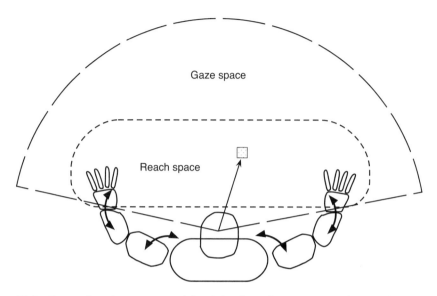

13.3 The reach and gaze spaces of the iCub robot. These spaces can be coordinated through the behavior known as *hand-regard*.

depends upon the maturity of the vision system, measured by how well the gaze space has developed. Similar thresholds apply when lifting the constraints on the hands and torso, based on the levels of hand-eye coordination achieved.

The region of space that the arms can move within is known as the *reach space*, and it defines the limit of reach of the hands. Figure 13.3 shows a downward view of the local workspace of the robot and indicates the reach space of the hands and the gaze space that can be seen by the eyes. To reach for interesting objects, or indeed to operate at all in the work area, locations in these two spaces must be coordinated in some way. But not only do these spaces vary in size, they are also completely different in their structures and frames of reference. The gaze space is centered in the head, so the eye targets are seen in terms of radius and angle of the line of sight (known as a *polar coordinate frame*), while the hands move along various locally curved paths according to the joints activated in the arms. This mismatch is one of the major problems facing our robot—just as it is for babies.

OBJECT PERCEPTION

Just as with humans, a robot cannot make much sense of the raw pixel data received from a camera. In the visual cortex, we perform some preprocessing of the incoming retinal data, and we need to provide a similar function for our robot. The visual signal from the central area of the image is processed to extract four basic features: brightness, color, movement, and edges. Objects usually have well-bounded areas of similar visual properties (that's how we distinguish them from other objects), so a cluster of pixels with almost the same feature value is called a *blob*. Blobs are simple to detect for color and brightness; as for movement, the feature detector has to notice a group of pixels all changing value at once. The fourth feature detector looks for contrast edges; that is, a group that has a division between brighter values on one side and darker on the other (exactly as we saw in chapter 6, figure 6.1).

The blobs produced from these four features are tested for correlation. For example, if the brightness blob has approximately the same size as the movement blob, then this is evidence that part of an object is being seen. If the color blob is also similar, and/or if the edge finder finds edges that fit around the blob boundary, then this is taken as even higher confidence that an object fragment has been found. The degree of correlation is a measure of salience, and if it is high enough, the blob details are stored in a memory for proto-objects.

EXPERIMENT 1—LONGITUDINAL DEVELOPMENT

Following the constraint lifting algorithm, the iCub was started at the "birth" condition and allowed to progress. At first, only the eyes are active (and even if other actions occurred, they would be ignored owing to the poor sensory resolution at that stage). Then head and eye coordination was learned, followed by arm movements. Table 13.2 describes the observed activity during the experiment and a time-lapse video of a half-hour experiment (speeded up fourfold) is available.[2]

The most obvious result was the emergence of very different qualitative behaviors. This was noticeable even within stages. For example, during saccade learning, three behavior types were apparent; at first, the

Table 13.2 Behavior development on the iCub.

Behavior	Description	Time first appearing (min)	Duration (min)	Duration of stage (min)
Saccading	Accurate, direct moves to peripheral stimuli	3	17	20
Gazing	Coordinated eye and head moves to fixate on stimuli	35	25	40
Arm movements	Hand regard—gazes at hand and reaches to visual targets	62	58	60
Torso movements	Moves at waist while having an accurate gaze at objects	125	15	20
Repeated touching	Reaches out to touch over and over	140	N/A	20
Pointing	Points to objects out of reach	160	N/A	10
Object play	Explores object affordances	173	27	40
Stacks objects	Places one object on another	210	N/A	20
Learning ends	Experiment ends	230		

Note: Typical early experimental data. Modified from table VI, Law et al. (2014b), with permission from the Institute of Electrical and Electronics Engineers (IEEE).

babbling covered large sweeps across the retina, then most saccades approached near the foveal region and needed a few corrective moves, and finally, one single accurate saccade was produced (most of the time).

When the arms first became active, a rather ballistic motor babbling characteristic of young infants were observed (Piek and Carman, 1994). This changed into hand-regard behavior when vision was sufficiently mature to recognize the hand in different places. Hand regard is important for controlling the arms and coordinating their reach space with the gaze space. When these spaces are correlated, in relation R_{arm}, it becomes possible for the hand and eye to signal attention about their respective coordinates when an interesting object or event has been detected, either by vision or by touch.

The arms and head are coupled at the shoulders, so once these have been coordinated, any movement of the torso will shift the gaze space together with the arms and thus not interfere with the eye-head-arm relationships already established. The relation between torso movement and its effects on the gaze space is learned as R_{torso}. The efficacy of this ordering is another justification for the proximal-to-distal constraint release idea. Torso movements also expand the reach space, creating more opportunities for reaching. A phase of repeated touching of objects then ensued, during which these sensorimotor experiences were consolidated and remembered.

Interestingly, pointing behavior emerges from the mismatch of the gaze and reach spaces. When a pattern for reaching for objects within the reach space has been successfully learned, the robot will attempt to reach for objects outside the reach space. Normally, this will fail, and the action will be downgraded for that location and eventually discarded. But the failure behavior looks like a pointing action, and if the experimenter moves the object nearer, it may then succeed, and this will be recorded as a two-step acquisition of the object and be available for use again.

The duration times of the stages reflect the fact that eye movements are more rapid than head movements and that arm movements are slower than both. It is important to note that stages are not fixed boundaries and behaviors will intermingle across stages. For example, while arm movements are being learned, eye coordination will still be improving and refining the visual space. The experiment occurred in real time, and much data was logged during all the actions and events. This remarkable development sequence, produced in less than four hours on a real robot, can be partly attributed to the bootstrapping effect of each stage creating a basis for the growth of later behaviors.

EXPERIMENT 2—THE GENERATION OF PLAY BEHAVIOR

When global excitation is low (meaning that there is a lack of novel stimuli), our developmental algorithm triggers the start of play activity. However, there is no need for a specific "play generator" module; there is no special implementation of play behavior. It is much simpler than that—play is produced as a natural function of the associative memory of past action experiences. When play starts, the memory extracts a previous experience that best matches the current context, while allowing some variation on the action. In this way, a known action can be tried in a new context. Various possibilities then arise: Either the action fails, another event distracts attention and previous memories are executed, or the play process might continue and produce a sequence of actions, some of which lead to new action situations.

We can illustrate this process by considering a given set of actions. Consider the following five actions: The Reach action simply moves the hand at the end of an arm from one location to another—from X to Y. Another move is a (horizontal) Push action, which can occur when the hand is in contact with an object and the object then slides with the hand along the surface of the worktable without grasping. Another action, Press, was provided as a primitive action in which the wrist makes a downward hand movement to press buttons during experiments. Figure 13.4 shows the iCub play bench. This was designed so that if the robot reached up to the space above a button and then Press was executed, the hand would tilt to perform the action.[3] Finally, we have two basic operations, Grasp (closing the hand), and Release (opening the hand).

Now, with the hand at location X, consider the consequences of each of the five possible actions. Figure 13.5 shows the actions as rectangles and the states of the world before or after an action as ovals. We see that sometimes actions can have no useful effect at all—Grasp and Release make no sense if there is no object involved, and Press and Push change things only if there is something to change. Even a previously successful action can sometimes fail; the brackets indicate components that have been experienced before, but with uncertainty—they don't always occur. The hand location is always reliable because proprioception is an internal sense and is not affected by the environment.

When a play situation arises, the memory of past actions is accessed, so in theory, any one of the five actions might be attempted. However, our associative memory also uses excitation and other data about previous action to match the candidate memories to the current context. Hence, a Reach to another location is most probable in this situation, but if an object was initially at the same location too, then Grasp or Push would be more likely.

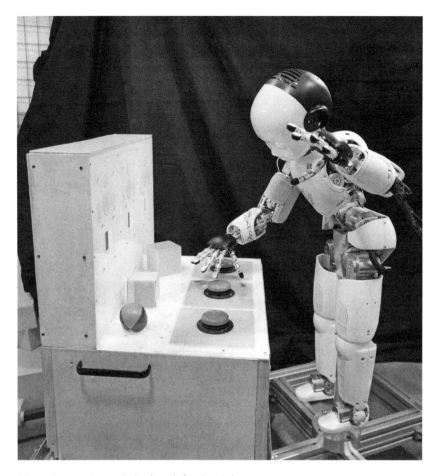

13.4 An experimental play bench for the iCub.

During one of the experiments, the system discovered a very unexpected new action—the use of an object as a tool. Figure 13.6 shows some action sequences. The far right path shows the normal preconditions for Press; the hand is at an object location and also touching the object (in this case, a button). However, during the session, the robot followed the left path; it reached to an object, picked it up, took it to the button, and then pressed the object down, thus pressing the button.

What happened here was that the associative memory suggested trying the Press action because "touching" and "holding" are very closely related conditions. (Touching events are tactile finger events that change

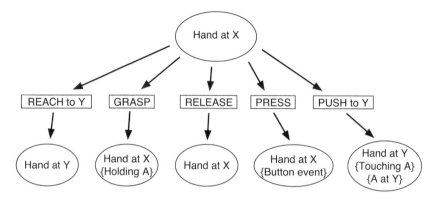

13.5 Possible consequences of actions. The brackets indicate things that may or may not occur.

into holding events when grasping is sensed by the proprioceptive sensors on the fingers.) The button was triggered, so Press was successful, and a new memory, a new skill, was learned. Further examples of tool-use behaviors, such as using objects to reach beyond the reach space boundary, have been explored in other experiments. The central path in figure 13.6 is another example of how play can generate "useful" compound actions; in this case, a series of Pushes can bring a group of objects together at one location. It is the subtle design of the associative memory and the resultant matching of prior experiences to new situations that provide the internal motivation for the production of new skills to deal with new possibilities. Statistics on the occurrence of events, objects, and memories are logged and used to increase the likelihood of trying familiar or successful items, while ignoring or even discarding those that do not get used.

The iCub system described here plays out sensorimotor events on a substrate of learned sensorimotor relationships. It pays attention to novel or recent stimuli, repeats successful behavior, and detects when reasonable competence at a level has been achieved. Successful experience is remembered, and intrinsically motivated play discovers new behaviors and generalizes from previous patterns. An early infancy expert conjectured: "Gradual removal of constraint could account for qualitative change in behaviour without structural change" (Tronick, 1972).

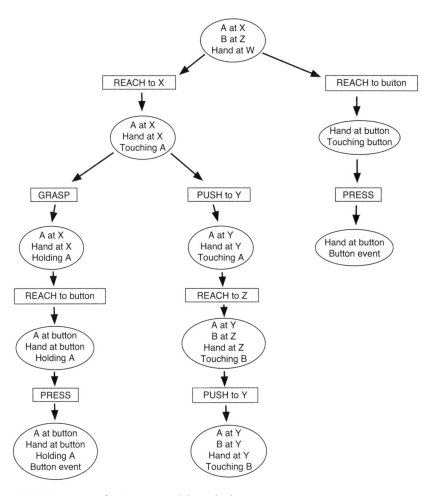

13.6 Sequences of actions created through play.

HOW DOES IT WORK?

Figure 13.7 shows the sensorimotor architecture underlying the experiments described in this chapter. It consists of motor subsystems for the eyes, neck, torso, arms, and hands; the imaging space of the retina; a memory for potential object fragments; a general memory of action experiences; and the gaze space—a short-term spatial memory of the local environment. All the iCub sensory inputs and motor signals are received and generated, respectively, on two-dimensional (2D) arrays of elements

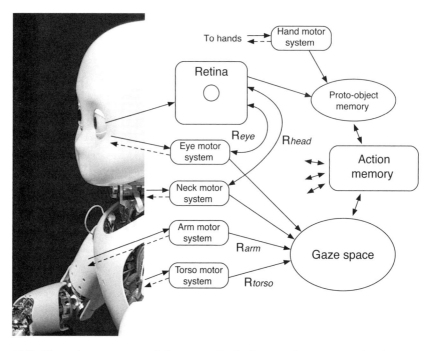

13.7 The main components of the system. Dashed arrows denote motor signals.

called *maps*. Each motor subsystem contains a map that captures the relationship between the motor actions and the sensed proprioceptive feedback. These maps supply the locations of the various body parts to the gaze map, thus forming an egocentric body-space representation (Shaw, Law, and Lee, 2015). The complex connections among these components have to be learned through the relationships shown in figure 13.7. The proto-object memory combines visual features with grasp sensing to build evidence of object experience. The action memory stores details of actions and their context within the sensed environment.

Our mapping method, as used throughout, takes inspiration from the brain. The human cortex is a multilayered, 2D sheet, folded into the familiar scrunched shape seen in textbooks. Although only around 3 mm thick, the cortex covers an area of about 1,400 cm^2.[4] All the sensory and motor systems are mapped out in various regions on this cortical sheet. These maps are topographic; that is, local structure is preserved—two stimuli that are close in one map also will be close in connected

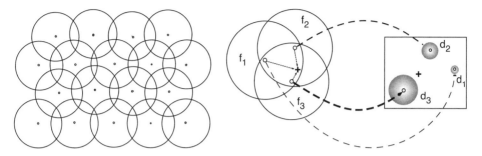

13.8 Left: Overlapping fields, shown with low overlap for clarity. Right: Accurate localization, with only three fields.

maps (Kaas, 1997). It is extraordinary that this uniform, repeated structure supports all of the wide variety of cognitive activity that goes on in human minds.

The elements in our maps are not independent, contiguous cells, like the pixels in cameras and vision systems; rather, they are circular regions (called *fields*) that overlap with their neighbors (see the left side of figure 13.8). Overlapping fields have the advantage that the location of a stimulus can be measured with much greater accuracy than is possible by simply noting the location of a single pixel in a conventional contiguous array. This is because a point stimulus, as in the right side of figure 13.8, will activate several sensors to a degree proportionate to its nearness to the field centers, and a simple averaging process can then reconstruct the stimulus's location in another map.[5] This effect is known as *superresolution* in vision, or *hyperacuity* for other sensory modalities (Westheimer, 2012). Research into robotic tactile and somatic sensing is increasingly looking toward hyperacuity. For example, tactile skin sensors spaced apart with a resolution of 4 mm have produced a hyperacuity of 0.12 mm, a 35-fold improvement of localization acuity (Lepora et al., 2015).

Overlap provides real gains when transferring data from one map to another. For any given level of error, the number of fields required to cater to that error reduces as overlap increases.[6] For example, an array of 100 by 100 units can be covered with an error rate of 1 percent with only 50 large fields, rather than 10,000 pixels for a conventional image. This means that only 50 learning steps are required to build the complete structure. We have used overlapping arrays extensively and have investigated many arrangements, from predefined structures to dynamically created arrays of elements, and from uniform arrays to random placements (Lee, Law, and Hülse, 2013). This report also shows how maps can stretch and distort to capture non-linearities. For our detailed results on maps and mappings, see Earland, Lee, Shaw, and Law, 2014.

The brain hosts multiple maps, often representing the same area at different scales (Derdikman and Moser, 2010). Our early experiments on different field sizes and coarse-to-fine effects suggest that hierarchical maps may be very useful for trading accuracy for speed. The receptive field size of visual neurons in young infants is reported to decrease with age, which leads to more selective responses (Westermann and Mareschal, 2004).[7]

We also looked at different proprioceptive coding schemes and the effects of variations in excitation parameters. Such variables can alter the order of local actions; but the general behavioral patterns and the coordination structures built across all the experiments were remarkably consistent (Lee, Law, and Hülse, 2013).

A series of experiments were designed to probe the role of the constraints and constraint lifting ordering. The effects of different ordering of arm motor action were explored and these confirmed the effectiveness of the proximal to distal ordering in overcoming the 7D-to-3D kinetic problem (Law et al., 2014a, 2014b).[8]

Interestingly, it is almost impossible to learn both the relation between retina and eye movements and between retina and head movements at the same time. This is because the two variables of the retina are then influenced by four motor values—another classic intractable problem, just like the arm control problem. However, if one relation is learned first, then things become easier. We found that the maps didn't have to be completely finished before using both together; so long as one had a good head start in the learning, progress could be made. See Law, Shaw, and Lee (2013) and Shaw, Law, and Lee (2014) for details of experiments that explore the successful learning of the gaze space.

We use Barlow's "suspicious coincidences" as a major guiding principle for perception. The proto-object memory relies on this to find visual blobs that consistently suggest object characteristics. Basic preprocessing can identify high-confidence blobs in an image, and the proto-object

Table 13.3 Heuristics for building pairs of visual patches.

Pair relation	Patch-1 and patch-2
Appeared together	At same time, not necessary but supporting evidence
Disappeared together	Also not necessary but supporting evidence
Moved together	At the same time—strong evidence
Moved similar	Same distance and/or direction (additional evidence)
Contiguous	Closeness of patch-1 and patch-2 (additional evidence)
Seen before	Count of appearances of this pair together

memory then tries to establish if they group together in ways that could represent an object. From studies on human object perception, Spelke (1990) argued that the visual patches that make up objects obey certain rules. Our system implements some of these rules or heuristics.[9]

Objects may have size, shape and color, but these are secondary features for young infants. It's all about movement; if you want to attract a baby's attention, wave, shake, or flash a bright object. Things that do not move at all are usually ignored. So most of the heuristics are based on movement (as seen in table 13.3). Each discernible blob that reaches a confidence threshold is recorded as a "patch" with various parameters, including the precise time of appearance and disappearance.

Any two patches that have high evidence of being related are linked and stored in the proto-object memory. The links are graded for strength of evidence and a simple counter records how often a pair of patches has appeared together. As the system sees more views of objects, so individual patches become linked together as pairs, and then linked into dense clusters of patches. The theory is that dense clusters are likely to represent the component elements comprising a single object. A cluster is considered to be a proto-object if (1) it has strong links between the patches in the cluster, (2) it has weak links to patches outside the cluster, and (3) most of the patches move together most of the time. Thus, static patches and patches that move in random ways, or are only seen once, are quickly ignored. Proto-objects can act as a means for generalizing, comparing, and recognizing similar objects. For detailed results of visual and proprioceptive sensory resolution, including the effects of changing visual maturity on proto-object perception, see Giagkos et al. (2017) and (Braud, Giagkos, Shaw, Lee, and Shen, 2017).

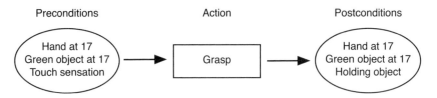

13.9 Schemas: the format of action experiences stored in memory.

An action memory is needed to recall past actions and match them to current needs. We take inspiration from the idea of a mental schema: a notion that captures the context and results of actions. The schema concept was used extensively by the Swiss psychologist Jean Piaget, and it has a long history. A schema can be simply defined as a memory element consisting of a precondition part, an action part, and a postcondition part. Figure 13.9 shows an example of a grasp schema. The preconditions require that the hand is at the same location as an object and is also touching it. If the grasp action is performed, then this memory expects that the hand ends up holding the object. There have been several software implementations of the schema concept; Drescher (1991) is well known, but earlier work by Becker (1973) and Plummer (1970, 1972) are more relevant.

These action memories capture the experience surrounding an action and can be used in different ways. Note that any of the three components may be empty, or ignored. Consider each of these in turn: If a precondition is known and an action is being considered, then the memory can give an answer to "What will be the outcome of this action?" Or we can work backward and ask what preconditions need to be achieved to reach a desired outcome from a given action. Or, if both preconditions and postconditions are known, then the schema memory can be asked, "What action could cause this change?" An associative memory provides all of these very different results.

The main processes of the action memory include creating new schemas, selecting a best match schema, producing generalized schemas from several examples, and chaining schemas for formulating and executing multipart actions. A generalization mechanism looks at sets of similar schemas and attempts to produce a single general schema at a more abstract level. This becomes a powerful way of moving from the specific (e.g., location) to the general (e.g., how to move an object anywhere). Schema chaining, as seen in figure 13.6, allows for cases in which the feedback of an action isn't immediate but still needs to be recognized as being useful as a potential postcondition or goal.

The dynamic nature of our schema system, allowing major changes during actions, during generalization, and during schema chaining, is a significant difference between our system and other implementations of the schema idea. This description has been necessarily simplified; there are many other features and subtleties in the design; see Sheldon and Lee (2011) for more background, or for the fine detail of the implementation, from the architect himself, read Sheldon (2012). For more on play behavior, see Kumar et al. (2018) and Lee (2011).

There has always been interest in training robots through imitation, reinforcement, and other forms of shaping (e.g., Dorigo and Colombetti, 1998), but few previous studies actually involved development or psychological issues. This is changing; today, staged development is regularly on the research agenda (see Ugur, Nagai, Sahin, and Oztop, 2015, for a useful overview). Also, the synthesis approach is now more in evidence (Kuniyoshi et al., 2007; Prince, Helder, and Hollich, 2005). A range of robotic hardware is used in developmental robotics, while iCub work often uses simulators (Tikhanoff, Cangelosi, and Metta, 2011).

Many implementations use artificial neural networks (ANNs) of the kind described in chapter 9. Our models have some neural-like features (such as layers of maps with overlapping fields), but these are built at a higher level of abstraction than neuronal modules and are fully accessible for investigation and analysis. Another framework for implementation is based on dynamical systems theory, and several leading developmental psychologists have followed this path (e.g., Lewis, 2000; Smith and Thelen, 2003).[10] Our approach offers an alternative to these two methodologies—a third way.

The appendix summarizes the insights, principles, and lessons learned from our research work. Chapter 14 continues with future paths for these models, including communication, self-awareness, and applications.

# III

WHERE DO WE GO FROM HERE?

14

HOW DEVELOPMENTAL ROBOTS WILL DEVELOP

The simplest meanings are in fact not simple at all. They rely on remarkable cognitive mapping capacities, immense arrays of intricately prestructured knowledge, and exceptional on-line creativity. They also rely on the impressive, and poorly understood, human ability to resolve massive underspecification at lightning speeds.
—Gilles Fauconnier, *Mappings in Thought and Language* (1997, 187)

In part I of this book, we looked at the basic ideas behind robotics and artificial intelligence (AI) and saw how AI influences so much modern technology. We found that AI is very powerful and is set to continue to be a major technology in the modern world. AI has been a great success, both scientifically and commercially. However, we also found that AI systems and applications, almost without exception, do not function in a way that is similar to human intelligence or behavior. This is not a criticism—it matters little to the success of AI whether systems operate in the same way as humans; it is simply a fact that we don't yet have any advanced working models of human thinking and behavior with anywhere near the success of existing AI systems.

In part II, I suggested how this imbalance between AI and machine models of human intelligence might be addressed. Instead of building cognitive systems, I suggested growing them through a process of synthesis and repeated experimental refinement. This developmental approach is heavily dependent on scientific knowledge across a range of human centered sciences, with a pivotal role for developmental psychology. Developing robots in this way is a slow process that requires considerable resources—both skilled people and expensive equipment. Many experiments are needed, and each small step in infant or human development has to be considered and possibly implemented. But the gains are cumulative, and we are building up knowledge and scientific understanding that will fuel more rapid advances.

In this part of the book, we look toward the future and consider the implications of developmental robotics, advanced AI, and the wider issues. Clearly, some of this will be speculative and can only be based on current trends and past history. However, it is important to avoid wild predictions that cannot be properly justified; I will stick to likely short-term developments based on evidence and reasoning. I'll also make a few remarks about the longer term, mainly about the risks of hyperbolic journalism and poorly justified futurism. The last chapters will broaden out to cover technology in general: we must always consider the context of our plans and developments.

HOW DEVELOPMENTAL ROBOTS BEHAVE

First, let's start with our iCub robot system from part II, and consider where this research path will lead. Experiments on real robot systems such as our iCub produce results at multiple levels.

At the implementation level, we tried various detailed designs and technical options and evaluated and learned better methods to use in the future. Examples of these results include the mapping technique, the value of overlapping fields, the excitation mechanism, and the fine-tuning of the enaction algorithms. The main effect of different implementation variations is to influence the efficiency and performance of the system; they usually do not preclude alternative designs being tried.

Other results deal with the major model features, such as the role of constraints and staged development in the LCAS model, the value of the "start small" principles, and the role of the schema memory system, all discussed in Part II. These are all significant results, in that changes to these model features can have major disruptive effects on behavior. If these elements are not right, then the system will not operate well, if at all. Some of these results are quite unexpected and emerge from the experience of running the experiments (e.g., the central role of proprioception and its dominant relation to vision).

At another level, there are results concerning the general behavior of the system: Does it follow the timelines proposed in the preparatory work? Do the actions appear compatible with human behavior? Does the learning follow reasonable growth trajectories? This is the level that is most engaging to the interested public: the general observed behavior and its immediately apparent animal-like (or not) qualities. It is difficult to predict future innovation at the detailed implementation levels, but we can say something about the general behavior that we should expect from the next step toward advanced developmental robots.

At this point, our robot has reached somewhere around five months on the infant timeline. It started from "birth" with some basic prenatal tactile sensing and random reflex actions. It has learned to control its eyes to perform saccades and fixate on visual stimuli, learned to incorporate head movements into its eye saccades, gradually gained control over its arm to reach a hand out for a visual target, and learned the effects of torso movements and combined them with reaching actions. Visual properties of objects, such as movement, color, and brightness, have been integrated with other object experiences, such as hand-grasping proprioception and finger tactile patterns, to form proto-object structures. The schema memory has collected a range of primitive action patterns and is used to generate tentative actions in response to novel stimuli. This long-term memory creates new schemas for successful activity, generalizes about other schemas and builds up a large repertoire of sensory-motor experience.

In general, the robot has built up a good egocentric body map of its peripersonal space, in which it has coordinated and integrated hand and eye behavior. It also has a crude understanding of objects as entities with distinct visuohaptic experiences (experiences that combine visual, tactile, and proprioceptive features), and it can manipulate them (so far as its limited mechanics will allow). The novelty criterion has proven effective as an intrinsic driver for action at a time when goal-directed behavior is meaningless, and the algorithm for babbling or play has provided a framework for discovering new action sequences that are appropriate for new tasks.

Every stage relies on all that has gone before, so our plans for further development will build upon these current competencies. Further hand activity is required to find new and different ways to grasp differently shaped and sized objects. These are known as *affordances*, and the considerable body of work on grasping in both humans and robots will be valuable here (Borghi, Flumini, Natraj, and Wheaton, 2012).

Similar refinement is needed to better discriminate static objects as more of the physics of object behavior are discovered; this includes gravity (falling and rolling objects), solidity and elasticity, and weight and mass effects. All of these can be experienced through play. More exploration and understanding of animate objects will also reveal their independent behavior, and this will be captured in new schemas. An action schema can be matched with both preconditions and postconditions even if the action is not compatible. So, for instance, a "pushing" schema [Object X at A, move arm to B, Object X at B] could be a close match to the autonomous movement of X [Object X at A, null event, Object X at B]. Here, the robot notices a change that is the same as one that it recognizes, but that did not involve a robot action. This has strong resonances with the "mirror neurons" in the brain, whereby we recognize the actions of others.

These groups of cortical neurons fire when we perform a particular action like picking up a tool, but they also fire in exactly the same way when we watch someone else make the same movements.[1] This suggests a mechanism by which imitation leads to new skills: attempting to repeat actions made by other agents. Mirror neurons were first observed in monkey brains but were confirmed in humans in 2010 (Keysers and Gazzola, 2010).

Other developments include the emergence of gestures for communicating intention. We have already done work on this and shown how pointing may emerge from failed reaching behavior. An out-of-reach object initially stimulates reaching attempts that fail, so any schemas generated from this will tend to decay and not be used. But if an observer responds to the outstretched arm by moving the object closer to the robot, the system then builds a schema that initiates pointing as the first step in a two-step process of obtaining out-of-reach stimuli. If this schema is successfully repeated, it becomes established and appears in the robot's behavioral repertoire as imperative pointing. Other gestures are a fascinating possibility—prelinguistic babies are capable of displaying considerable communication skills through gesture (Goldin-Meadow, 1999); try a few gestures when you next meet a baby, and you'll be amazed at what they understand!

Other work has shown how a single spoken word can be attached by association to an object or action. If the robot hears a sound pattern every time a certain object is present or a particular action is performed, then that sound can be linked to the object or action. The schema system takes care of this using a device known as *associated observation*: a sensory event that is not a conditional part of the schema. There are often several features that could be candidates for an association, but an attachment is not established until repeated experiences of the event have disentangled the context. The sound could be a bell or a noise, but we have used an operator speaking single words into a microphone, such as "grasp," "drop," "red," "blue," "ball," or "block." These become associated as verbs or nouns according to whether they were said repeatedly with an action or with an object. Later, repeating known words provokes associated schemas and, effectively, commands can be performed: "Grasp red ball." There is no language structure involved in this early work; these are simple, one-word tokens. Hence, "Ball grasp red" would have exactly the same effect.

One important development is the growth of purposeful or goal-directed behavior. Play is essentially goal-free activity—at least, it is free from external goals. But purposeful actions are clearly evident during infant play. These emerge as the infant gains control over its actions, which then become learned as skills. As more internal goals become established, they can be used when trigger conditions arise. During play, new paths to known outcomes are discovered and the better path (which is shorter or more desirable in some way) will take precedence. Extrinsic goals relate to basic biological drives, such as the relief of hunger, pain, or physical discomfort, or the attainment of some desired system state. We have not used extrinsic goals yet in our work, but the attachment of tokens to actions and objects does show how external goals can be specified by other agents during interactions.

An obvious limitation of our particular iCub is that it is immobile; the robot has legs, but they do not have sufficient hardware functionality to allow it to walk.[2] This means that the totality of its experience has taken place within an egocentric space, a space defined by the limits of vision and reach. Some events and stimuli outside this egosphere may be noticed, but there is no direct experience or action relating to them and therefore they will have no meaning. Newborn infants start like this but as they begin to move around, by crawling and then walking, they carry their egocentric space to new locations. Eventually, they build a new space, an allocentric space, which is related to the ego reference frame but extends to include external, more remote features. The local egosphere still exists,[3] but it is combined with a space that contains external structures and alignments, such as "North," or paths or boundaries. An allocentric model of space is necessary for the navigation through and exploration of spaces that are bigger than the agent itself.[4]

Although our work is ongoing and there is still much to do, we have gained data and experience, and we have designs for future experiments. Everything described here is feasible, and indeed has been demonstrated in various research experiments across the developmental robotics community. Of course, for any one project, there are many infant features and abilities that are not even partially modeled, but a broad range of developmental aspects are being explored. Many things remain to be refined and enhanced, but we can be confident that this new paradigm will deliver learning robots that develop and grow in a way quite unlike AI robots. So let's now consider what such a robot may be able to do in the future. How will such robots behave?

In the following sections, I describe expectations for developmental robots in the near future, but I am not going to go beyond the scientific facts. All the developments I suggest have already been proven in one way or another, whether in research work or studies in related fields. For example, the hands on our iCub robot are quite advanced, with good finger control and tactile feedback but, like most robot hands, they have their own particular limitations compared to the human hand. Nevertheless, active research on robot manipulation has developed some very sophisticated hands.[5] It is reasonable to envisage the next generation of humanoid robots as having superior manipulation capabilities, skin-based tactile sensing, and dynamic responses. By pulling together the best results from a range of work across the research community, we can build a picture of the next stage of advanced developmental robots.

TAUGHT, NOT PROGRAMMED

The most striking difference between developmental robots and AI robots is that the former cannot be programmed to carry out specific tasks. The idea of directly changing things inside the robot's "brain" goes completely against the philosophy of the developmental paradigm and is also impractical and ultimately counterproductive.

Developmental robots only gain meaning through action-based experiences shaped by the anatomy and previous history of that particular robot. This means that if we want to program a robot to carry out desired tasks, then we must train it in the real world. Thus, training becomes a form of customizing in situ. We must show the robot the objects we wish it to manipulate and suggest, through gesture and by example, the actions we require the robot to perform. This is similar to the educational practice known as *scaffolding*: designing environments that hide distractions and emphasize the desired features, so that training has a sharp focus. This process is neither as random nor as tedious as it might appear; the system has learned names for previous objects and verbs for well-known actions, and these can be used to simplify the training. The mirror-neuron function facilitates the imitation process of learning from observed action; showing a new action can allow it to be perceived and then repeated and rehearsed a few times.

In this way, previous tasks can be used as a basis for learning new tasks. The repertoire of nouns and verbs will increase in time, and it can be used to form goal statements. Specific external goals could be incorporated, such as a *hunger* variable for low energy levels, but it is a matter for further research as to how external value systems might be combined with intrinsic motivations.[6] For the moment, though, we can ignore external motivation and see how far intrinsic motivation algorithms take us.

All of this will be in real time—necessarily so, because the robot must carry out its own actions during training. Completely new tasks will require some exploratory play time to learn actions relevant to new objects and situations. This is another major difference from AI methods. Developmental systems have very limited numbers of trials for learning, unlike AI systems that can be trained by presenting large batches of sample data in repeated (usually offline) cycles until the correct response has been attained. This separation of learning from behaving would be a major violation of our approach, as discussed in part II.

Another remarkable consequence of the developmental approach is the individuality of the robots. All robots might be identical when they are manufactured, but as soon as they start work in different situations, their trajectories of experience, and therefore their competencies, gradually diverge. They become specialized by their owners, their environments, or both. Just consider a couple of mass-produced home-help robots; after being in (widely) different homes with (widely) individual users, they will develop very different perceptions, experiences, and skill sets. Unlike the unrealized ideal of the personal computer, personal robots really will become customized to their owners' needs.

This introduces difficulties for maintenance. As each robot effectively has its own history, embedded in its long-term memory, it is no use generalizing when trying to diagnose faults (as we do effectively for other machines, such as cars). There will be a demand for sophisticated diagnostic tools, but the most effective method may be to correct the problem in the same way as it was trained—by programs of corrective scaffolding. An undesirable behavior can be suppressed by training on an acceptable variant that replaces it. The robot service center of the future might offer a home-based retraining package, or in severe cases, the robot might be brought to a remedial school. When learning by imitation, it may prove more convenient for robots to learn from other robots rather than people (perhaps requiring an especially trained corps of instructor robots?).

Skills can be transferred by dumping a developed robot brain into an empty new body. The schema memory would be the main item to be copied, and this would be a trivial operation. In this way, best practice can be spread; many robots can share the best skill set. But there is a big snag: The robot hardware must be identical in order for this transfer to work; if there are any differences (e.g., in limb length, sensing resolution, or motor power), then there will be problems because the whole life experience is based on the enactive approach: brain and body are integrated. So individuality of behavior across identical bodies can be transferred, but any anatomical or hardware changes prevent transfer.

For example, the skills learned on our iCub could be transferred to another iCub, but not to another make of humanoid robot (say, a Meka robot).[7] Similarly, major hardware upgrades, such as adding another camera or arm, could not be introduced and keep the existing development; the whole system would need developing from scratch again, with the new addition in place. This is the same for humans; a great deal of time is spent readapting to either bionic additions or biological losses.[8]

Fortunately, we can examine the whole brain of a working robot. We can look at the signals and data flowing inside the robot, as well as recording external behavior and interactions. By seeing through the robot's eyes (subjective viewpoint) and observing the behavior (objective viewpoint) at the same time, we get a "god-eye view" that is a powerful scientific tool. These systems thus become test beds for a range of investigations, trying out ideas about enactive cognition for robots, as well as psychological hypotheses.

KNOWING ONESELF AND OTHER AGENTS

In any exchange with another person, we have to understand the meaning of what they say to us—not just their words, but also the intentions of the person playing a role in the conversation. We build up a picture of why the other person is saying certain things and what she or he is like. So we appreciate other people as agents, similar to ourselves, with needs and wants, which helps us to understand their actions and utterances. In order to do this, we maintain internal models of people so we can interpret and predict the meanings inherent in our interactions. To some extent, this is a necessary ability for any agent that needs to communicate with others, including human-friendly robots.

To do this, it is important to know one's own behavior and role as an agent. This *self-model* can then be used as an approximation when other agents are encountered. I can thus interpret someone else as a human, with similar expectations and needs to my own, but I will have to make adjustments to my model of this person as I learn about his or her differences and particular behaviors, foibles, desires, and motivations. Our developing iCub has the elements of a body model: it knows the spatial properties of its arms and torso and how to move in its peripersonal space; it knows the sensory spaces and their relationships, and it is continuously building up knowledge of objects and the effects it can create. However, a model of the self for communication purposes is much more complex than this; every little detail of an agent's behavior can be important and contain meaning. So, how can we build such complex models?

Actually, models of extremely complex systems are being created every day. The flow of liquids in pipes, the activity of electricity in circuits, and the behavior of hot gases inside internal combustion engines and jet turbines all can be simulated so authentically that computer models often replace the real hardware in much of the analysis and testing work. This is often known as *virtual prototyping* and is a consequence of the reductionist approach to science. The mathematics involved is able to capture the behavior of the components of a system and calculate how they interact together.

Unfortunately, reductionism often breaks down when humans are components in the system, mainly because we don't have much understanding of their behavior. Economists, financial experts, urban planners, and market analysts all build models of complex social systems, but they suffer from unreliable results and intense internal debate about the nature and details of the models.

What are the desirable characteristics of models of other humans? They will be quite different from the numerical equations in most of science. They must be very flexible, be able to hold tentative properties and structures, and accept approximate or coarse information; and they will be created for and identified with individuals, while allowing generalization across common traits and features. This sounds like a tall order, but potential groundwork already exists in the field, known as *soft computing*. This includes techniques such as fuzzy, qualitative, and analogical reasoning, all of which explore human cognition and reasoning characteristics. Model-based reasoning has a long history, and there is an abundance of cognitive model-building expertise.[9] For our needs, we can draw on this work to create an incremental, flexible model, with constant testing against experience, detection of discrepancies, and internal diagnosis and repair to ensure compatibility with current behavior and interactions. Very relevant to this will be experimental psychological research into how we infer feelings and intentions from others (Busso et al., 2004).

SELF-AWARENESS IS COMMON IN ANIMALS

If we have a good understanding of ourselves (i.e., a self-model) and can use it to make sense of our actions and experiences, then we can say that we are *self-aware*. In the past (pre-Darwin), this has been thought of as a defining feature of the human species. But research into animal cognition has opened up a more enlightened view of our place in nature. We know that dolphins and octopi are as intelligent as humans, but other animals are cleverer than we thought. For example, sheep are often considered as pretty unintelligent animals; they seem unable to form any kind of purpose, never mind a plan. If you have ever tried to communicate with or help a sheep, you probably came away with the impression that they must have a very limited brain. However, they are actually not stupid. Research has shown they can recognize the faces of other sheep (which we find hard to do) and discriminate among different facial expressions. Sheep seem to be able to recognize people, and tame sheep have long been known to respond when you call their names. They can also discriminate among colors and shapes and perform as well as lab monkeys in certain cognitive tests (Morton and Avanzo, 2011). They also navigate by forming lasting memories of their surrounding environments.

Other surprisingly smart animals include certain birds. The intelligence of jackdaws, magpies, and other corvids is well known. They are innovative, can solve puzzles to get food, have good memories, and can use tools. Self-awareness is demonstrated by the mirror test; these birds recognize themselves rather than treating their reflection as another bird. It is interesting that this particular bird family has brain-to-body mass ratios similar to that of the great apes; and the chicks have very long growth periods, during which they learn from their parents.

Perhaps you knew about the clever crows, but even more remarkable is recent research on the common domestic chicken. Chickens certainly don't have the brains of the crow family, but they are smarter than we think. A recent study reports: "Chickens can demonstrate self-control and self-assessment, and these capacities may indicate self-awareness. Chickens communicate in complex ways, including through referential communication, which may depend upon some level of self-awareness and the ability to take the perspective of another animal. Chickens have the capacity to reason and make simple logical inferences. They are behaviourally sophisticated, discriminating among individuals, and learning socially in complex ways that are similar to humans" (Marino, 2017, 141).

Going even further down the biological tree, we know that honeybees solve navigation problems and communicate new routes to others, but it seems that even the humble fruit fly can tell us about cognition. This laboratory workhorse is widely used for genetic studies, but it has been discovered that fruit flies have similar biochemical signaling processes to those found in mammals. It's suggested that because all brains are made up of networks of neurons, the range of different ways of using the neural, electrical, and chemical signals is limited, and so similarities across all organisms with brains are more likely than might be expected.[10]

It seems that the various mental functions and abilities that we associate with human cognition are present to some degree in many animals. It may be just a matter of degree that gives us our extended abilities to plan, reason, and construct new innovations; perhaps there is a critical level of cognitive components at which full-blown imagination becomes supportable.

ROBOT SELVES

The point of these animal examples is to illustrate that the basic machinery for a simple form of functioning self-awareness may not be that complex. A body model is an essential requirement, along with adequate perception of how one's actions affect the body and change the entities in the environment. We have the technology to build good models of physical systems, but we also have modeling expertise in less well defined areas, particularly the human sciences, involving social, behavioral, and cognitive fields. Soft computing methods have explored human mental models and attempt to capture the essence of the influences and interactions that occur in human behavior. I believe that these methods could provide the tools we need to support a simple level of self-awareness. Our iCub robot has already constructed a body model, and the sensory, motor, and memory components are fully integrated into the internal, and externally perceived, system states.

Once the self-model has reached sufficient competence to display some self-awareness, it should then be possible to run a duplicate model (or, more likely, a partial model), as a tentative, malleable reference for another agent. The idiosyncrasies of the perceived agent would shape the model, which would be stored for use in two versions: one as a developing understanding of that particular agent (identified by a name or other token), and the other as a contribution toward a general agent stereotype that compiles commonalities across agent models.

As we are interested in creating useful robots, and not trying to model complete simulations of human states and behavior, we do not need to build emotional or human values into our robots. Of course, the ability to detect emotions is necessary for there to be empathy and rapport in human-robot relations; and such aspects of robotics are being actively researched under the topic of affective robotics (Kirby, Forlizzi, and Simmons, 2010). But the ability to display emotions needs a bit more control.

Emotional behavior is application specific; it relates to context. I do not think it is necessary in the general case to be able to generate a full range of human responses; in fact, an angry emotional robot could be quite scary (!) and an impartial, factual, calm demeanor is more suitable for most purposes.

Important inspiration for modeling the self also comes from eminent neuroscientists: "The same neural structures involved in the unconscious modelling of our acting body in space also contribute to our awareness of the lived body and of the objects that the world contains. Neuro-scientific research also shows that there are neural mechanisms mediating between the multi-level personal experience we entertain of our lived body, and the implicit certainties we simultaneously hold about others. Such personal and body-related experiential knowledge enables us to understand the actions performed by others, and to directly decode the emotions and sensations they experience. A common functional mechanism is at the basis of both body awareness and basic forms of social understanding: embodied simulation" (Gallese, 2005, 23).

CONSCIOUSNESS

Although I argue for self-awareness, I do not believe that we need to worry about consciousness. There seems to be an obsession with robot consciousness in the media, but why start at the most difficult, most extreme end of the problem? We can learn a lot from building really interesting robots with sentience, by which I mean being self-aware, as with many animals. The sentience of animals varies over a wide range, and it seems very unlikely that consciousness is binary—either you have it or you don't.

It's much more probable that there is a spectrum of awareness, from the simplest animals up to the great apes and humans. This is in line with evolutionary theory; apparently sudden advances can be traced to gradual change serendipitously exploited in a new context. As I've indicated, there are many animal forms of perception and self-awareness, and these offer fascinating potential. Let's first try to build some interesting robots without consciousness and see how far we get.

Support for this view comes from biophilosopher Peter Godfrey-Smith, who studies biology with a particular interest in the evolutionary development of the mind in animals.[11] He traces the emergence of intelligence from the earliest sea creatures and argues for gradual increases of self-awareness. He says, "Sentience comes before consciousness" (Godfrey-Smith, 2017, 79) and claims that knowing what it feels like to be an animal does not require consciousness. It seems entirely logical that we can replace the word *animal* with *robot* in the last sentence. Godfrey-Smith also argues that "language is not *the* medium of complex thought" (2017, 140–148, Italics in original), which supports the view that symbolic processing is not a sufficient framework for intelligence.

In any case, it is important to recognize that the big issues in human life—birth, sex, and death—have no meaning for robots. They may know about these concepts as facts about humans, but they are meaningless for nonbiological machines. This seems to be overlooked in many of the predictions for future robots; systems that are not alive cannot appreciate the experience of life, and simulations will always be crude approximations. This is not necessarily a disadvantage: A robot should destroy itself without hesitation if it will save a human life because to it, death is a meaningless concept. Indeed, its memory chips can be salvaged from the wreckage and installed inside a new body, and off it will go again.

Consequently, such robots do not need to reason philosophically about their own existence, purpose, or ambitions (another part of consciousness). Such profound human concerns are as meaningless to a robot as they are to a fish or a cat. Being human entails experiencing and understanding the big life events of living systems (and some small ones as well), and human experience cannot be generated through nonhuman agents. If this contention is accepted, it should counter much of the concern about future threats from robots and superintelligence.

Two Nobel laureates, Gerald Edelman and Francis Crick, both changed direction following their prize-winning careers. Edelman won the prize for his work on antibodies and the immune system, and Crick was the codiscoverer (with James Watson) of the structure of the DNA molecule. Both started research into consciousness as a second career. Edelman experimented with robots driven by novel competing artificial neural systems (Edelman, 1992), and Crick looked for the seat of consciousness in the brain (Crick, 1994). They didn't agree on their respective approaches, but their work, well into retirement, produced interesting popular books and showed how fascinating the whole topic of consciousness is. Despite their mutual criticism, their general goal was the same: They both thought that the circular feedback paths in the brain somehow supported consciousness, and they were looking for structural mechanisms in the brain.

I have already argued that sentient agents, like robots, need not be conscious, but they must be self-aware. In any case, it is a reasonable scientific position to start with experiments with models of self, self-awareness, and awareness of others and see how far the results take autonomous agents. Then the requirement for, or the role of, consciousness, can be assessed by the lack of it. This is not a structural approach, based directly on brain science as with Edelman and Crick, but rather a functional approach: What do models of self offer? How do they work? What is gained by self-awareness? What is missing from the behavior of sentient robots that consciousness could address?

COMMUNICATION

Early infant behavior includes two sorts of interaction: social, with parents and others; and physical, with objects and artifacts in the local environment. Play with physical objects supports the learning of space, causality, object properties and affordances, and the way the world/environment behaves. Social interactions are essential for the development of communication. The first social interactions occur as the infant builds up competence in following the caregiver's eyes. Face detection begins very soon after birth and enables eye contact to be established. Interactions usually take turns, with the caregiver initially following the infant. After around three months, the baby can indicate her or his feelings and the caregiver can follow the baby's interests more closely. At about nine months, the infant gains more understanding of the caregiver's intentions and the interactions become more symmetrical.

After this, an important milestone is reached, in which both can follow the other's gaze to a common target; this is known as *joint attention*. Pointing is also used to indicate and attract attention toward target objects, and vocalizations are commonly involved. Eventually, interactions become fully symmetrical, with the infant able to both follow the caregiver's attention and also direct the caregiver to its own interests. Eye contact and joint attention are key competencies because they correlate the attention of the participants in the interaction. Both agents know they are dealing with the same target and thus have a kind of empathy toward the target. Any words, symbols, or sounds that occur during joint attention can be associated with the target, so names or other identifiers can be established and used. As ever, timing is important; the agents must be synchronized within very tight intervals.

We have demonstrated in our experiments how verbal tokens can be attached to objects being noticed, and this suggests a framework for learning some basic language use. We have also shown how failed reaching attempts can transform into pointing gestures. The next stage is to incorporate face detection to support synchronized interactions, leading to proper joint attention. This should be combined with affective computing methods to enable the detection, learning, and recognition of interacting humans' emotional responses and states, such as pleasure, surprise, and frustration. By carefully following the findings of relevant psychological research, we can model and utilize similar learning behaviors in our developing robots.

DEVELOPMENTAL CHARACTERISTICS

What are the main features of this approach to robotics? Developmental robots will be able to reason about their own actions; they'll be self-aware to that extent. They will be aware of other agents' actions and compare them with their own: a primitive theory of mind. They'll be rational, they'll know facts that they believe to be true, and they will have sets of beliefs that they know are not strictly true but might be justified through future experience (and thus become facts eventually). They will also have episodic knowledge of sequential events and experiences, as well as various rules and heuristics that they have learned for dealing with their environment and its challenges. But they won't have emotions, in the sense that they won't become angry or display fear. Of course, they will be able to show their current state; for example, when they keep trying to achieve something that repeatedly fails, they might give facial expressions for frustration, but they won't act on it and become distressed or violent.

The duplication of human thinking patterns is too ambitious an achievement even for developing robots. Robots of the future will always be robots; they will be better than the zombies we have at present, but they certainly won't be confused with humans. The cold and calculating demeanor so long associated with computers will not be completely banished because accurate answers will be available, and expected. Humanlike errors will not occur, as reliability will be considered important. These robots will have the potential for unlimited patience, which clearly would have to be restricted by some mechanism, probably an excitation/boredom process, as we have used previously. But these characteristics should be seen as a benefit rather than a disadvantage. With no intrinsic bias or distractions, there will be no muddled thinking, and less doubt derived from emotional stress, so more productive decisions and behavior should result.

The intrinsic motivation to explore the world, play with novel situations, and be open-ended in approaching new experiences and skill acquisition will produce exciting and interesting new behaviors, well beyond current fixed-goal motivations. Robots will have individuality, with their own preferences and abilities (depending on their prior lives), and the models of self and others should provide a rich research area to engage scientists for some time.

WILL ALL THIS HAPPEN?

In case you are beginning to think that these projections for developing robots are too speculative, too much like science fiction, and without proper support, it's worth noting that all the areas mentioned are being actively investigated by current researchers. For example, the concept of body maps and the perception of local body space are now very active areas in robotics. Conferences regularly focus on the results of research on models of local space; see Hoffmann et al. (2010) for a review of this area. For self-awareness, there is a range of different takes on this, including Chatila et al. (2017); Blum, Winfield, and Hafner (2018); Gallagher (2000); and Rochat and Striano (2000). The disproportionate spaces allocated to different body areas in the brain have inspired a number of models of multimodal mappings between proprioception and tactile and visual space to provide an integrated sense of egocentric space (Matin and Li, 1995).

This leads to investigations into how such self-referential spaces can extend and relate to the larger allocentric space required to navigate, to understand relations between different objects, and, indeed, to relate to other individuals. See Klatzky (1998) for an overview, and Snyder, Grieve, Brotchie, and Andersen (1998) for some neural inspiration.

The importance of action in shaping early perception, and indeed the development of the mind, is well founded (Hunnius and Bekkering, 2014; Bushnell and Boudreau, 1993). And to learn more about infants' ability to distinguish the purposeful actions of other agents, see Woodward (2009) and Gerson and Woodward (2014). It is always the target of the action, rather than the means, that infants use to infer the intentions of others.

The overlapping mapping technique used in our work is inspired by the mappings widely found in the brain, as reported from neuroscience. These are effective in supporting the growth of low-level sensorimotor processes, but we also found they continue to be useful for proto-object recognition and the coordination of cross-modal activity. Remarkably, almost the same mechanisms are advocated by language theorists for meaning construction, the mental constructions produced during language processing and perception. Gilles Fauconnier says that "mappings between domains are at the heart of the unique human facility of producing, transferring, and processing meaning" (1997, 1). He presents mapping methods for modeling human reasoning, analogy, and creativity. It would be tremendous if it turned out that this simple mathematical concept contained the basic substrate process for all human cognition, from the lowest levels of stimuli processing to the most abstract levels of social and creative thought.

As we saw in chapter 10, there is a lot of activity in AI around the issue of how internal knowledge and experience should best be represented. Autonomous intelligent agents will need to find ways of incorporating new experiential knowledge into their existing frameworks. This is sometimes known as *learning-to-learn*; finding a good way to learn speeds up actual learning. For example, Tenenbaum's group has produced systems that learn structures of many different forms and discover which form is best for given data (Lake, Lawrence, and Tenenbaum, 2018). These structures are graphs (trees, arrays, orderings, etc.) and are similar to the mappings produced by our approach. This work aims at similar goals to ours[12] but lacks embodiment—physics simulators are used, despite criticism (Davis and Marcus, 2016)—and it does not incorporate several key features of development, including autonomy and intrinsic motivation.

Another important topic is play. Among developmental psychologists, there are few who doubt the profound necessity of play behavior for healthy human development. This view also extends into neuroscience (e.g., Pellis and Pellis, 2009). The idea of play and nonspecific exploration has been examined in several robotic scenarios, such as in the work of Steels for language acquisition (Steels, 2015) and in Oudeyer's lab, where Sony AIBO robots discovered actions and contingencies in a playground environment, Oudeyer, Baranes, and Kaplan, (2013). This work also employs developmental constraints and shows them to be effective in directing growth. Research into motivation is also exploring emergent tool-use (Forestier and Oudeyer, 2016). Der and Martius (2011) describe a series of experiments on play involving embodiment, self organization, and homeokinesis that produce self-motivated, self-exploratory behavior. A wide range of robot morphologies were used, and new emergent behaviors evolved. Unfortunately, they are not interested in developmental robotics or cognitive development and instead concentrate on the kinematic possibilities of skill learning.

So there is a great deal of relevant research under way. From a practical viewpoint, the work is fragmented and needs integrating. Longitudinal experiments are one way of advancing this.

WE MUST GET OUT MORE ...

Edwin Hutchins (1995) in his book *Cognition in the Wild*, examines how people think and respond to real-life situations and crises. He argues that most of what is known about cognition is learned in the laboratory, and this ignores the sociocultural environment in which we all operate. He gives examples where groups of people think their way out of nasty problems: large-scale crises that require complex cooperation and that no individual could solve on their own.

It is proved that humans behave differently in laboratories compared with natural situations, and we must be aware of this fact when using psychological data. Much of the background information used to build and justify AI systems is drawn from such sources; for example, conversational systems are often scripted, either directly or by processing patterns from large data collections. Robotics has the advantage that experiments are carried out in the real world, thus giving the scope for the freedom that Hutchins describes; however, vigilance is necessary to resist the creeping intrusion of shortcuts and simplifying constraints and stay true to the science of development. Hopefully, developing robots will be able to get out (of the lab) much more in the future!

OBSERVATIONS

- Children are not incomplete or partial adults; their early stages of development in infanthood are finely tuned for learning skills and experiences that will be very valuable to them in adult life. This is a fundamental lesson for developmental robotics.
- Consequently, developmental learning takes place in situ and in real time. Tasks and goals are learned through play or taught through demonstration and imitation, not by programming.
- Almost by definition, developing robots will be individuals. Each will have different experiences and learning as they follow their own "life" paths.
- Self-models are important. They comprise maps of the body, mappings of local space, and repertoires of actions and outcomes, intentions and goals. They allow the agent to understand, assimilate, and control its own behavior.
- Models of other agents can be created by copying a self-model and then modifying it through the ongoing experience of the other agents.
- Sentience is necessary, but consciousness is not. Basic self-awareness and the awareness of others are necessary abilities for social robots and will be a sufficiently challenging research area for the foreseeable future. Full consciousness is interesting but is a diversion at this stage. Advanced developmental robots do not need full consciousness.
- We cannot expect robots to duplicate all human characteristics. They will never become artificial humans because they will not experience all human life. As for all AI machines, they won't have any concept of life, death, and survival, at least for themselves. This means that strong emotions will not be important to them, and they will be more platonic than humans.

15

HOW AI AND AI-ROBOTS ARE DEVELOPING

I never think of the future. It comes soon enough.
—Albert Einstein, translated interview by David P. Sentner, *Clearfield Progress*, December 1930

The decade starting in 2010 has been a flourishing period for robotics, in terms of both technological advances and penetration into public life and awareness. We have already outlined a possible future path for developmental robots, but how are AI-based robots likely to advance? Some of the difficulties of applying traditional artificial intelligence (AI) in robotics were discussed in part I, so let's now be more positive and look forward to developments that we might reasonably expect in the near future. Although the topic is robotics, AI is so important for robotics that we must look at AI developments too.

TASK-BASED AI

Any AI that has a discernible target application or function can be seen as task-based. I've argued that all existing AI, commercial and research, falls into this category. Some AI researchers will have qualms about this position, saying that a task-based view is too limiting, too restrictive. They argue that solving one problem, achieving a task, can be a step toward knowing how to solve many more—every advance is an increase in our understanding of cognitive processes and has scientific value. I agree entirely; AI has generated an enormous body of theoretical and experimental results that enlighten and advance the scientific understanding of computing and information processing.

Recent developments in neural networks, probabilistic reasoning, machine learning (ML), and data mining have established the maturity of AI as an absolutely central component of computer science and modern software practice. But the task-based categorization is not intended to diminish AI—it is just a way of organizing our thoughts so that we can focus on the concrete results and the down-to-earth analysis of what AI systems actually do, what they don't do, and what they can't do for robotics.

One effective way of assessing progress is by looking at the results of the many international challenges and tournaments. These competitions cover a huge range of topics and are aimed at both robotics and AI. They usually offer big money prizes for the winners and focus on specific, just-out-of-reach tasks that inspire innovation and creative solutions. Challenges are often repeated annually, so progress is easily quantified by measuring the yearly improvements in the results.

Several significant challenge events have already been discussed, and some of these (and others) are listed in table 15.1. For robotics, the challenges tend to fall into a few main categories: industrial applications, domestic situations, rescue missions, and "fun and games." Perhaps the best-known robotics challenge is the Defense Advanced Research Projects Agency (DARPA) autonomous driving challenge series, mentioned in chapter 2. The off-road events (especially in 2005) gave a major confidence boost to the self-drive research community and relevant interested companies. Unfortunately, DARPA's humanoid robot follow-up to this success was a flop and stands as an example of ambitious overreach for the technology of the time. It was sad to see the climax of this massively funded program involved sophisticated robots collapsing while struggling with a common task like opening a door.[1]

Industrial robotics includes fine assembly work, bin-filling and picking, and advanced mobile functions. Rescue and emergency robotics have become more active since the 2011 nuclear accident in Fukushima, Japan, and often involve land, marine, and flying robots working together. The "fun and games" challenges include robot soccer, which has been going since 1997 with the original aim of producing a robot world champion soccer team by 2050. The RoboCup organization now also includes robot leagues for domestic, rescue, and industrial challenges. Service and domestic robotics challenges are usually based around indoor situations with specific tasks, such as retrieving objects in domestic settings or delivering items in complex buildings.

Challenges are an extremely cost-effective way of stimulating innovation, a fact noticed by the many companies and government agencies that sponsor them. Many of the best software minds find an ingenious technical challenge irresistible, especially if there is access to expensive hardware, free software, and the chance of large financial rewards. The scale of investment can be seen from the $3.5 million from DARPA (for their 2013–2015 series), to the largest innovation program in civilian robotics in the world, the European Robotics Innovation Program (SPARC; see www.eu-robotics.net/sparc), with a total fund of over 2 billion euros (2014–2020, with 700 million euros from the European Commission). The message here is that if you want a problem solved, produce a well-defined goal and set up a challenge or competition centered around that goal. We are likely to see many more challenge events, and correspondingly more human records being broken.

HUMAN-LEVEL AI (HLAI)

The term *Human-Level AI (HLAI)* refers to AI systems that have reached, or exceeded, some level (records or milestones) set by human performance. The fanfares and accolades break out when an AI program achieves world-class performance and beats the best human exponent.

We have seen a great deal of HLAI in recent years, with many records broken. Table 15.1 shows several challenges, marked with a $\sqrt{}$, that have met or exceeded human levels of performance. This usually means that the challenge is not run anymore; reaching near-perfection removes the motivation to reduce errors any further. Also, once the human world champion has been beaten, there is no way to measure further improvement, and any future challenges are often between competing AI systems. It's noticeable from table 15.1 that more success is being achieved in AI than in robotics.

Table 15.1 Some challenge competitions in robotics and AI.

Topic	Organizer	Status
Robotics		
Off-road self-driving	DARPA Grand Challenge 2004, 2005	✓
Urban self-driving	DARPA Grand Challenge, 2007	✓
Land robot trials	European Land Robotics, from 2006	
Rescue robotics	DARPA Robotics Challenge, 2013–2015	
Emergency robotics	ERL	
Industrial, bin-picking	Amazon, from 2015	
Industrial	ERL	
Service robots	ERL Service Robots	
Domestic/social robotics	RoboCup@Home	
Volkswagen Group	Deep Learning and Robotics Challenge, 2017	
Robot soccer	RoboCup, from 1997	
AI		
Games–chess	World Computer Chess Championship	✓
Games–Go	Computer Go, UEC Cup, Tokyo	✓
Games–others	Computer Olympiad, from 1989	
Conversation (chatbots)	Loebner Prize (Turing test) from 2006	
Language processing	Many contests (e.g., Machine Translation, EMNLP 2017)	✓
Netflix	Best user prediction algorithm ($1 million prize)	✓
Face recognition	Face Recognition Challenge, IARPA, from 2017	
Recognition of images	ILSVRC, from 2010	✓
Handwriting recognition	Document Analysis/Recognition (ICDAR) from 1991	✓
Representation learning	ICLR, Learning Representations, from 2013	
ANNs	NeurIPS, from 1987, competitions on classification and other topics	
Data mining	Kaggle, 19 current competitions, from 2010	
Coding competitions	ACM Collegiate Programming Contest, from 1970	

Status key: ✓ indicates results very near or exceeding human-level performance, suggesting that future challenges would be unnecessary.

Abbreviations: ACM = Association for Computing Machinery
EMNLP = Conference on Empirical Methods in Natural Language Processing
ERL = European Robotics League
IARPA = Intelligence Advanced Research Projects Activity
ICDAR = International Conference on Document Analysis and Recognition
ICLR = International Conference on Learning Representations
ILSVRC = ImageNet Large-Scale Visual Recognition Challenge
NeurIPS = Conference on Neural Information-Processing Systems
UEC = University of Electro-Communications in Tokyo

But human performance is a relative variable with many surprises. Human face recognition has now reached a very impressive level, with deep methods feeding into all kinds of applications: phones, cars, cameras, personal media, data searching, and security. Performance has exceeded human error rates in many cases. Despite this, AI techniques cannot compete with *superrecognizers*—people who, after looking at a single photograph, can identify faces in a crowd from all sorts of angles and in all kinds of lighting conditions. Less than 1 percent of people have this ability, and they are highly valued by security services (Davis, Lander, Evans, and Jansari, 2016). No one knows how this works, or why the rest of us can't do this at all.

We can expect a lot more human-level benchmark challenges in the near term. New challenge areas will demolish other human levels of performance as different tasks are attacked. However, by definition, all these will be task-based systems: A challenge has to have a target! Some AI proponents seem to believe that Artificial General Intelligence (AGI) will emerge, or become possible somehow, when a sufficiently large number of HLAI benchmarks have been surpassed. As we saw in chapter 10, the conference series "Human-Level AI" promotes this view. The 2019 joint multiconference included "AGI 2019: the only major conference series devoted wholly and specifically to the creation of AI systems possessing general intelligence at the human level and ultimately beyond." But general AI and human-level performance are orthogonal concepts; there is no need to connect them. A theoretical AGI system might surpass humans in many areas but lag in others; it is its generality that matters. Conversely, we may have many systems, each of which exceeds human levels (as we do already!) but have no idea of how they might be combined into a general system that could go on to conquer other problem areas.

DEEP AI

The Deep Learning movement is certain to continue to grow and support many new AI developments. The pace of progress in this area has produced dramatic results in a very short time scale. The full implications and understanding of Deep Learning on multilayered neural networks are still at an early stage, but that does not stop the rollout of new products;

science lags technology in this area. Furthermore, everyone can now have a try at Deep Learning; you don't need access to a supercomputer anymore because new tools and hardware are opening up deep methods for everyone. Powerful hardware for desktop computers can be bought from companies like Nvidia[2] for a few hundred dollars, which works out at just a dollar or two per Graphical Processing Unit (GPU) chip. Further, free open-source software tools like TensorFlow[3] offer libraries for handling arrays of numerical data structured in deep layers. It is possible that Google will sell its powerful Tensor Processing Unit (TPU) hardware for running TensorFlow, or another company may produce something similar. If you don't want to be bothered with the hardware, you also have the option of running your deep system in the cloud, effectively renting some time and space on a massive computer server (probably a Google TPU farm).

I mentioned the ambitions of the Deep Learning researchers to find ways of extending neural methods to cover the traditional AI techniques like symbolic reasoning, planning, and testing and diagnosing complex systems. It's not yet clear if such sequential, logical processes make any sense on a distributed neural system, but if these goals were achieved, then this could be a potential threat to other AI research. Just as artificial neural networks (ANNs) have recently overtaken most of the older, nonneural computer vision methods, the rest of task-based AI might be outperformed too. If this happens, then the neural model might provide a platform for fully integrated AI systems. It is likely that the developers of such an integrated ANN will make some claims for AGI performance, but, for the reasons we have seen, this will still not be enough; it will be engineered toward tasks, and it will certainly not exhibit humanlike thinking or behavior.

Two big snags remain for nearly all neural net applications. These are closely related: verification and validation, as mentioned in chapter 9; and the black-box problem. For safety-critical systems in areas such as finance, energy distribution, security, and all high-integrity systems (and this extends deeply into the fabric of business and society), it is hard to see how a Deep Learning application could be awarded a reliability guarantee (never mind a safety certificate), given the stringent testing and authentication regulations required. Next, if black-box systems cannot explain their reasoning and justify their results, then they will not engender trust and will be rejected in favor of human-related services.

ROBOT DEVELOPMENTS

In terms of hardware, quite a few long-standing robotic difficulties—mechanics, dynamics, materials, power efficiency—have been overcome, and this has led to low-cost devices reaching the public for the first time. These often have limitations, such as plastic bodies, low-precision sensors and motors, and unsophisticated software; but they have opened up the range of social, ethical, and moral issues surrounding the prospect of widespread robotics to the public and politicians, which is surely a good thing. There is still quite a way to go with much-needed hardware advances. The research agenda includes creating soft robots that don't damage humans, with better tactile sensing (the sense of touch lags a long way behind artificial vision), more flexible skin with humanlike properties, and other biologically inspired improvements.

What does Deep Learning mean for robotics? There are many opportunities for big improvements in applications. For example, when it comes to manipulating objects with their "hands," robots aren't known for their dexterity. The kinematics, motor systems, and sensing capabilities of robot hands are far removed from human manual performance. For instance, see Jamone, Bernardino, and Santos-Victor (2016) for measurements of the grasping abilities of the iCub hand. This is a big issue for robotics, perhaps not surprisingly, because the hand is one of the most important and distinguishing features of humanity. Raymond Tallis (2003) argues that the dexterity and manipulative potential of the hand was responsible for the growth of self-consciousness and our unique sense of agency. Developing tool use laid the context for the growth of symbol systems and language.

A project called DexNet used the Big Data approach to train a robot on more than 6 million views of virtual objects (Mahler et al., 2017). The data was generated synthetically and used to train a convolutional neural network (CNN) that estimates grasp quality and predicts success for different grasp trajectories. When the robot then saw real objects, it could select an appropriate position for its two-fingered gripper to grasp and lift any object.[4] With a 98 percent success rate, this is certainly one way of improving dexterity and learning fine motor skills.

This kind of approach is known as *end-to-end learning*. For example, in grasping, the idea is to capture relationships between images of an object and the motor action required to reach and grasp that object. All the processes traditionally implemented between these ends (image processing, object detection, spatial learning, grasp learning, and error correction) are covered by the learning in one deep network. There are clear attractions in end-to-end learning, as all these stages do not have to be designed, analyzed, and programmed, but there are downsides too, such as lack of control of all the application nuances. One example of this is a project at Google Research, with Sergey Levine, who ran up to 14 robot/camera systems in a massive trial-and-error experiment to learn where to put the robot's two-fingered gripper to pick up an object. The robots ran for months, working on a range of several hundred different objects. This was a real brute-force approach (Levine, Pastor, Krizhevsky, and Quillen, 2016).

A review of Deep Learning in robotics is provided by Pierson and Gashler (2017). Reinforcement learning is a common component of Deep Learning systems and is becoming central to AI-based robotics; see Kober, Bagnell, and Peters (2013) for a review of reinforcement learning in robotics.

SOCIAL ROBOTS

Much of social robot research sees the main application in terms of helpers: assistants who provide support and care for those lacking full independence in some way. This leads to circumstances that define a given task to be achieved. For example, the European Robotics League (ERL) 2019 tournament specified subtasks such as "Getting to know my home," "Welcoming visitors," and "Catering for granny's comfort."[5] The robots were involved in activities like safe and autonomous navigation, grasping objects, recognizing faces, following people, understanding spoken commands and displaying facial expressions. While these are useful investigations, it is not obvious that a conventional robotic device is appropriate for home care. Issues of cost, complexity, and reliability can be more important than the actual functionality of the device, and the shape and appearance of a robot can be a deal-breaker as well. The requirements for social technologies extend far beyond robotics or AI, and they should be examined well before selecting them.

One area that will be essential is spoken conversation (including some textual and other language media). Virtual assistants have become very popular, with Siri (Apple), Google Assistant, Alexa (Amazon), and Cortana (Microsoft) even becoming household names.[6] These will continue to evolve and improve, but what will be involved in creating an effective social robot? How should an AI designer engineer a conversation; or, more precisely, how should they engineer the robot's part of a conversation with a human? There are various ways of approaching this. One way is to design templates for common patterns of speech. Another is to use Big Data and try to emulate previous similar responses. But, as seen in the Turing test, these methods are soon found to be inadequate. The real solution is a generative method that responds sensitively to the other person's pattern of speech or text.

The English alphabet has only 26 letters, but the number of words that can be assembled from these pieces is enormous; an average adult vocabulary is around 15,000 words. And the number of ways that words can be put together in sentences is practically infinite. (Actually, most random combinations of letters and words are meaningless, but this has little effect—a fraction of infinity is still infinite!) And the number of different sequences of sentences in a conversation is similarly enormous. So any conversation is likely to be unique, in the sense that some patterns, some combinations, will be different from everybody else's. Of course, there are lots of common patterns, where we all use the same idioms, the same greetings, the same motifs and stylistic effects, and so on. But in any nontrivial conversation, these will be of low value, as they do not enhance the information content appreciably.

For long-term companions, our conversations build from previous encounters, and we expect our companions to remember what we discussed previously and be able to pick up the threads and continue in a relevant way. This means that conversations are effectively long streams of information, building up cumulatively, and the meaning of each new sentence is related to the totality that has gone before. This is why a *generative* approach is so important. We have to create a sentence in response to our partner's utterances, and it has to have meaning; it cannot be purely factual data, like the question-and-answer systems we looked at earlier in this book. Episodic memory is important here, as events and sequences form the basis of many conversational exchanges.

The kind of information processed during conversations comes in very different categories. For example, information about matters in the world differs from information pertaining to the human in the conversation, and also from information that has meaning to the robot's personal experiences. Similarly, some things are factual statements that are true or false, and others are episodic information about processes and time-related sequencing. The references to these statements in speech has to be correctly parsed so that the robot knows which experience we are talking about (the robot's or the human's, or even somebody else's). There is a great deal going on here, all happening in real time—the next statement could change everything. It's impossible to precompile this into some kind of computable response for any social remark. This is truly one of the biggest AI challenges. While the Turing test may well be passed for 10 minutes, or even half an hour, during which the robot might fool a person into thinking that it was a human, that kind of simulation will be of no use for a companion robot that runs for months or years at a time.

ARTIFICIAL HUMAN INTELLIGENCE (AHI)

A natural question to ask is: How far can task-based AI reach toward human intelligence? Can it get close to Artificial Human Intelligence (AHI)? A useful way to look at the problem is to consider an actual scenario. Imagine a machine that is not performing properly, and we want to diagnose the issue. I was going to suggest a washing machine, but some people might simply buy a new one! So to make it a bit more serious, let's consider a plane coming in to land, but the pilot finds that the landing gear is not responding and the wheels won't come down. The pilot aborts the landing and flies around nearby while the ground staff try to figure out what the problem is and how to fix it. There are at least three ways of trying to diagnose it. First, we could scan the available data on all relevant aircraft and search for similar patterns (all aircraft have detailed logging files on maintenance, events, and performance). This would involve matching the type of plane and the symptoms observed by the pilot, and hopefully finding some record of a similar case and what was done to correct or bypass the fault. This is a Big Data approach.

A second and more sophisticated method is to look up the engineering manuals and follow any diagnostic processes as prescribed. This is a form of rule-based reasoning: The manuals have been compiled by skilled engineers and contain knowledge of the functioning of the complex machinery, both for normal and faulty operations. By following the rules for diagnosis, a picture of the problem may be defined that then indicates a solution. This is the knowledge-engineering approach, as exemplified by systems like Cyc that try to capture and exploit both commonsense knowledge and specific technical knowledge.

Finally, if these methods failed to help (and we are all getting anxious by now—or at least the people on the plane are!) some senior engineers might be called in from the manufacturers. These would be design and development engineers with intimate knowledge of the landing gear and associated systems. They would be able to simulate the faulty state of the plane, either in their heads or by using software models, effectively building a model of the faulty system based on the symptoms; and by trying different options and running these forward, they might find the path to the best outcome.

This simulation approach is very powerful because it is generative; it allows completely new scenarios to be examined, and it can predict future behavior. The other methods will eventually run out of relevant data or guidance rules, but simulation models are limited only by their level of authenticity. For example, most diagnostic manuals only deal with one fault at a time, but if an aircraft sustains physical damage, both electrical faults and hydraulic faults may be present simultaneously. This can be modeled by inserting relevant "damage" modifications into the normal operating model to produce outputs that match the real system's behavior. The ultimate simulation is to use an exact copy of the thing itself—a classic example being the engineers at the National Aeronautics and Space Administration (NASA) using a second space shuttle on the ground, a complete copy in every respect.

The embodied approach to AHI, as described in part II of this book, offers the possibility of creating robotic systems that do not suffer from the issues we've exposed for AI-based robots. We want developing robots to learn from experience, like the aero-engineers, rather than just from books or rules. Our ongoing experiments show that we can make serious inroads and progress, and also that it is not as difficult as it may look. What AHI needs is more people, more work, and more results.

While AGI is seen by many as the ultimate goal of AI research, AHI has the goal of understanding social agency and autonomous development. It all depends upon what you are trying to achieve. As Steven Pinker said, "inventing the car did not involve duplicating the horse" (quoted in Brockman, 2015, 6). But trying to duplicate (perhaps parts of) the behavior of the horse is a very good way of understanding how the horse works and achieving some of its impressive abilities and actions.

OBSERVATIONS

- AI growth will continue into the future, with many new developments and applications, particularly using Deep Learning and Big Data. Many human-level benchmarks will be reached and exceeded, and records will be broken, but these task-based advances will not make AI any more general.
- Innovation in AI and robotics is greatly stimulated by competitions, challenges, and tournaments. These are big events that involve thousands of people, a huge range of topics, and some large cash prizes. They are compelling for technologists and are a low-cost way of funding potential breakthroughs.
- The integration of conventional (i.e., nonneural) AI modules into robots is fraught with difficulties and is unlikely to give quick and easy benefits.
- The "Deep and Big" style of AI will be applied to robotics and will produce some new and impressive results. The two main technologies involved will be deep neural networks and reinforcement learning. Although there will be many applications, particularly for industrial and service automation, their shortcomings will limit their use for autonomous social robots and safety-critical areas.

- Robot-human interaction is an essential part of future social robotics, and this is one of the most difficult AI areas. Research tends to partition social issues into subtasks, but fragmentation tends to finesse over the holistic issues. More integrated and longitudinal experiments are needed in this area.
- Social robotics emphasizes mental simulation as a key component of human interaction, but it is not as straightforward as the simulation of scientific models. A subjective viewpoint is also required. These features, involving models of the self and others, will eventually support generative processes that are needed for effective, real-time social interaction.
- Life and death is meaningless to a machine, and therefore advanced AI robots will not need to be concerned with their own survival. This removes the need for analogs of strong emotions, most of which have evolved from deep survival instincts.

16

UNDERSTANDING FUTURE TECHNOLOGY

Traditional scientific method has always been at the very best 20-20 hindsight. It's good for seeing where you've been. It's good for testing the truth of what you think you know, but it can't tell you where you ought to go.
—Robert M. Pirsig, *Zen and the Art of Motorcycle Maintenance*, 1974

The twenty-first century started out with people being quite anxious about future technology, and many worrying scenarios have been bandied about. Various new artificial intelligence (AI) technologies are sure to have an impact on our future lives, and there are some real threats on the horizon, particularly for employment and global economics. But it is absolutely essential that these technologies are not simply imposed on users, who have to find out the hard way whether they are really useful or if they are just passing novelties, time wasters, or even dangerous or damaging. It is vital that the general public (including potential users) has a basic understanding of the function and scope of such (possibly intrusive) technology. Only then can the implications be fully appreciated and opinions properly informed. This technology will often be complex; but complexity must not be used as an excuse for not providing information!

The review of basic AI in part I was offered to help this understanding. We must get everyone involved in order to encourage reasoned consensuses about acceptable levels of intrusion, security, and user control. This is the ethics of technology deployment. We all need to be involved in shaping the ethics of new technology and dealing with potential threats. This chapter and the next consider some of these issues. We start by discussing the rapid rate of change, and then consider the important matter of trust.

RAPID GROWTH—IT'S NOT REALLY EXPONENTIAL

Commenting on future possibilities for technology is always extremely risky. Any predictions are likely to become out of date by the time they are published. I have tried to avoid this by focusing on general principles rather than on specific products.[1] When new technologies are introduced or major breakthroughs unleash rapid developments, the take-up growth rate is often described as "doubling every n months" or "exponential." The data on the uptake of new products often does conform to such steep growth curves, but usually only at the beginning. Figure 16.1 shows a true exponential growth curve plotted along with a sigmoid (S-shaped) curve. Exponentials shoot up toward the vertical at some point (tending to infinite growth), but if some other moderating influence comes into

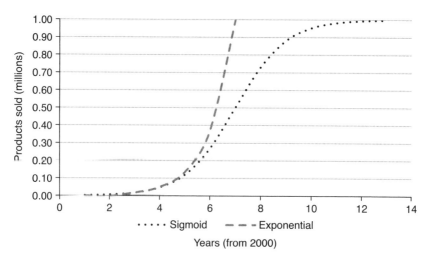

16.1 Is it really exponential? A typical plot of products sold per year.

play (as nearly always happens), then the growth rate will slow and may level off, as the sigmoid does, and eventually decline, as all products do. The important point is you can't tell what sort of curve you have at the beginning.

When smartphones were first introduced, the market was wide open, but when everyone has one (or two !), then manufacturers can sell more of them only if other phones are scrapped. If only one model were ever produced, then the smartphone growth curve would look like a sigmoid curve (and level off at the replacement rate for lost or broken phones). The manufacturers' ploy to stop the curve flattening, and keep them in business, is to make you discard your phone by bringing out ever better, supposedly must-have models.

The software engineering industry was established in the 1980–1990s and the demand for trained software engineers was fierce. At one point, there were so many vacancies for software engineers in the United States that some pundits plotted the data as an exponential curve and claimed that "nearly everyone in America will be a software engineer in the year 2000."

Another bizarre example: In the early 1970s, there was a good deal of concern about global population growth. The best computer models compared several global threats and found the most serious to be the burgeoning birth rate across the planet. Model projections "proved" that each person would have around 1 m^2 of the Earth's surface to stand on by 2000. Of course, that didn't happen, not because we'd have no room for food production and distribution well before that point, but because the birth rate actually began to decline (owing to improvements in education and health).[2]

Sometimes things really are exponential. The doubling of computer power roughly every two years is a remarkable feat, from which we have all greatly benefited. As transistors were squeezed ever more densely onto silicon chips, the computing speeds went up and, surprisingly, computers became much cheaper. We know the shape of the growth curve for silicon productivity because we have plenty of historical data, but even this curve is now slowing as the physical limits for maximum chip density are being reached. This illustrates the importance of knowing the causes and mechanisms behind the curves, and when the underlying situation changes or will change: Chip density has only recently approached the atomic limits, but global warming could become critical after only additional slight increases in temperature.

Exponential growth can't continue in the real world. Some other factor will always come into play before the growth variable continues far along the curve. This is because life is complicated! Birth rates and engineering don't depend upon single factors; rather, they depend upon a range of interrelated, interconnected influences and conditions. What may be a favorable condition for growth will change with time—situations always change!

GROWTH PATTERNS IN THE TWENTY-FIRST CENTURY—SO FAR

We are in a remarkable period for technology, especially in the areas of software and electronics. The range and volume of products sold have been enormous, often because they are opening up new applications and new markets. For example, the smartphone has been around for only a few years (the first Apple iPhone was launched in 2007), and yet we already can't imagine, or remember, life without it.

Robotics has smaller markets but has also seen burgeoning demand and rising sales. The automation of manufacturing accounts for many of the world's robots, but there are now many lower-cost commercial offerings for the education, domestic, and entertainment markets. The public interest in research robots like the walking Honda prototypes has encouraged a wave of more affordable humanoid designs (often composed of plastic rather than metal). Wheeled mobile platforms of many designs are available for service and support tasks, such as in hospitals, and two-armed upper-torso arrangements are now affordable for research and industrial experimentation and development.

The statistics for robot sales show increasing year-on-year growth (the typical rate is 15 percent per year). For industrial robotics (e.g., in manufacturing), the main demand comes from China, followed by South Korea, Japan, the United States, and Germany. The total number of robots sold is now almost twice the average levels seen before the 2008 crash—a remarkable recovery. China has seen the most rapid growth, with nearly a third of the global market.[3]

Looking at the recent acquisitions of robotics companies is also informative. The large corporations have been buying up smaller, specialist companies in increasing numbers. In 2016, more than 50 robotics companies were acquired, mostly by Chinese and American corporations. The average cost per company was over $500 million. In 2012, there were only nine new robotic start-ups, but in 2016, there were 128 new companies, with total funding of $1.95 billion. That's 50 percent more than in 2015. The biggest categories were unmanned aerial systems, agricultural robotics, service robotics for business and personal use, and self-driving systems.[4]

These figures will be superseded by the time this book is published, but they indicate the explosive growth that has occurred since about 2000, when the growth of commercial robotics and AI was slower and steadier. For robotics, the industrial and military sectors are predicted to grow to a combined global market of around $40 billion by 2025, but the big areas are the commercial and consumer sectors. The number of personal and domestic robots was up by 24 percent in 2016, and this is projected to continue at 30 percent for some years. A total of 40 percent of robotic start-ups have been in the consumer sector, which has recently been revised from $26 to $45 billion in combined markets. This reflects the penetration into more human centered applications and is due to remarkable financial investment,[5] falling prices of equipment, easier implementation of software and AI, and the stimulation of the self-driving industry.

Looking at the growth in AI companies, we see a very similar boom. In 2016, there were over 40 acquisitions of small, private AI companies, the main buyers being the large corporations such as Google, Intel, Apple, and IBM. These giants often bought several AI companies. Google bought the most: 11 over five years. During that time, more than 140 companies researching into novel AI techniques were acquired. Recent data shows that more than 3,600 AI start-ups have been created since 2013. These covered more than 70 countries, and their total funding amounts to $66 billion.[6]

The emergence of China as a major player in AI is remarkable for its suddenness and rate of growth. The world's biggest retailer, Alibaba, has greater sales and profits than all US retailers combined. Alibaba, which also owns Alipay (like PayPal, but bigger), is investing heavily in its already powerful AI and cloud services and is set to be the world leader in these technologies. Other huge corporations include Tencent (investments, web and phone services, and entertainment), Didi Chuxing (like Uber, but larger; it is the biggest ride-sharing company in the world), and Baidu (world-leading internet services, search engine, AI, image recognition, and self-driving systems). Baidu Brain is an artificial neural network (ANN) with highly accurate performance in voice and facial recognition.

All this very rapid growth has consequences. In the West, many start-ups, loaded with venture capital, begin quite small, often staffed by just a handful of enthusiasts, but most of them never grow to become medium-sized companies. They either fail or, if successful, are quickly bought up. This increases the power of the few very large corporations and reduces innovative competition. Another problem is the shortage of skilled staff, which is a familiar headache across the computing industry. Education and training schemes take time and cannot respond fast enough. Salaries rise as rich companies poach staff from universities and less well funded countries, thus creating a so-called brain drain and unbalancing the distribution of talent.

ARTIFICIAL GENERAL INTELLIGENCE (AGI)

It is interesting that quite a few start-up companies have begun work on Artificial General Intelligence (AGI). Despite the pessimism and lack of progress, there is a new feeling that if AGI could be realized, it would be transformative. The applications and commercial sphere would expand even more rapidly than today, and, essentially, a new industry would be created. This is why AGI has been called the "Holy Grail of AI" (Boden, 2016, 22).

Some of these companies are well funded; they look like regular businesses and have declared missions aligned with AGI. For example, Kindred, based in San Francisco, argues that humanlike intelligence requires a humanlike body; the company is working from that premise to build robots that understand their world.[7] Kindred's advisors include AI experts Richard Sutton and Yoshua Bengio.

A brain-based approach is being taken by Numenta, which "believe[s] that understanding how the neocortex works is the fastest path to machine intelligence."[8] Jeff Hawkins is a cofounder of Numenta, and its agenda of reverse engineering the neocortex closely follows his well-received views on AI and the brain (Hawkins and Blakeslee, 2007).

OpenAI is exploring "safe AGI" (with built-in ethical considerations) and has Elon Musk as a cofounder and cochair. It also has backing from Microsoft and Amazon and has committed funding of $1 billion.[9]

GoodAI, based in Prague, was founded by Marek Rosa.[10] The company is based around a detailed road map that ends in Stage 4, "super-human general AI." They take a long-term view and claim "not [to] be distracted by narrow AI approaches or short-term commercialization."

A mature approach seems to be that of Vicarious, which talks in terms of timescales of 40+ years and is very concerned about doing research before development.[11] Vicarious states that understanding is more important than chasing performance measured by standard benchmarks.

DEEP NETWORKS, LEARNING, AND AUTONOMOUS LEARNING

Clearly, Deep Learning has been the prominent boom technology of the twenty-teens. From recent breakthroughs showing that multilayered neural networks could solve difficult problems (whereas for decades, the accepted wisdom was that they worked only for a few layers), the performance from, and confidence in, these networks has soared.

Google and Microsoft both use Deep Learning for speech recognition over a range of inputs. Google and Baidu use Deep Learning for searching through photograph collections, visual databases, and other video documents and imports. Deep Learning is also widely used in language processing of various kinds: text analysis for searching and meaning extraction, machine translation, question answering, analysis of parts of speech, word sense disambiguation, and the list goes on and on. As we saw in chapter 10, deep networks are already into important areas like self-driving vehicles, and they are definitely going to take up a large portion of future AI technology.

But even more important is the way that AI is increasingly being used in various self-learning modes, whereby data patterns can generate programs that perform other functions. There also is a history of this in AI:

for example, self organizing systems, program generation by evolutionary algorithms, and self-evolving coding schemes. A classic example is Luc Steels's work on multiple agents that evolved segments of their own language through embodied interactions (Steels, 2015).

A recent case is Google's language translation system which, you may have noticed, suddenly improved its performance and now gives very good translations (between 10 of the most common language pairs). In 2016, Google switched from its previous Google Translate, a collection of systems for different languages, to the multilingual Google Neural Machine Translation (GNMT) system. This is multilingual in the sense that all translations go through the same intermediate stage. This allows the words and expressions for meanings to cluster together for many languages (Johnson et al., 2016). Headlines appeared such as "Google Translate AI Invents Its Own Language to Translate With," by publications that should know better.[12] But GNMT did not create its own universal interlingua. What happened was that it found a structure for clustering the commonalities that exist in languages expressing the same topics. This area is of much interest to linguists and psychologists who study these common connections between languages.

ARE THERE ANY DEAD CERTAINTIES?

Without projecting too far along the curve, we can identify a few areas of emerging technology that are fairly sure to become established. These can be indicated by signs of commitment in the industry such as significant new investments, favorable government policies, and major internal changes being made in production or recruitment.

For example, the signs are that electric, self-driving cars are almost certain to be widely adopted very soon. They were proved as a concept years ago, they have been under intensive development, they are now up to commercial standards, and literally millions of road miles have been covered in extensive testing. But the main reason is that the leading car manufacturers have nearly all decided that their future growth and profits are going to come from a switch from oil-based to electromotive power. After working with hybrids and other systems, they don't want to be left out of the race for electric vehicles and miss the many opportunities that will arise from this much more flexible power source.

A confirmation of this is seen in the increased activity in battery research and production. Tesla Motors and the global electronics company Panasonic have built a huge battery manufacturing plant in the United States. This plant, known as Gigafactory 1, started mass-producing lithium-ion batteries in 2016 and aims to produce half the world's supply of these batteries.[13] By initiating such a step change, the idea is to lower the cost of batteries (by 30 percent) and thus kick-start a new market. Batteries are one of the most expensive items in electric cars, so this, together with other manufacturers' innovations, will lower the cost of electric cars and make them even more attractive. This shows that the consumer doesn't always have a choice! When the car manufacturers can offer electric cars at considerably cheaper prices than internal combustion cars, then they force the market to shift. Of course, they benefit too—from massively reduced car manufacturing costs. We will all benefit from reduced pollution, increased efficiency, and lower costs.

Interestingly, the first self-driving car insurance policy was launched in the UK in 2016. This includes cover for installing software updates, navigation satellite outages, operating system faults, and hacking. This insurance assumes a person is always in the driver's seat. The vehicle is essentially seen as a super cruise control system (with lane assistance and other features), and the insurance is an extension of existing cover used for modern, gadget-packed, computer-rich vehicles.

But completely autonomous vehicles, with no one on board, will be an entirely different matter. As we saw in chapter 2, there are five levels of autonomy recognized for self-driving vehicles[14] and the kind of assistive automation available on the market is currently around levels 1 and 2. Level 4 vehicles (high automation) can drive themselves on most proper highways but still require a human driver on board, while level 5 vehicles (full automation), can self-drive in any kind of environment, don't need a human, and don't even need to have a steering wheel. Level 4 can be seen in the research vehicles being tested from all the major laboratories, but level 5 is still work in progress.

To see what a challenge full self-driving really is, imagine sending your own car to collect a friend from the airport, and on the way there, it has a crash with another, completely autonomous, unattended vehicle. Assuming that no third party caused the collision, then who pays for the damage? If there is no driver, then the liability for an accident must fall on either the car's owners or its manufacturers. The owners have a good argument (due to their not actually being in the car, never mind driving it!), and it is difficult to see the manufacturers accepting all the liability for the entire lifetime of the millions of cars they will produce. There may be a legal case that the manufacturer is to blame because it is their software that is acting as driver, but they are sure to resist this vigorously. Perhaps some form of standard accident tax or a national fund paid into by owners will emerge.

The legal and ethical issues surrounding completely autonomous cars raise complex difficulties over and above cars with controls. Indeed, the technical AI problems are much harder too. The extra step to cover all driving conditions rather reflects the difference between AI and AGI—it's much easier (even when it's hard) to design a system that can handle a (large) range of different conditions than it is to design a system that can handle *all* conditions. *All* means circumstances where everything and anything could happen.

When the majority of vehicles become self-driving, the number of accidents will decrease dramatically. There will be no speeding or traffic jams, policing costs will decrease, and road maintenance will be more controlled and efficient. There will always be people who like driving and want to drive themselves, but they will become conspicuous if they try to speed in the context of a road full of autonomous vehicles, so maybe there will be no "fun" in driving and that will accelerate the shift toward a totally autonomous traffic system. However, the transition period to this transport utopia is the difficult stage.

Another safe bet for new automation is the agriculture and horticulture industries. Harvesting crops is one of the most backbreaking, low-paid, manual jobs around. Not only does it offer little reward, but by its very nature, it tends to exploit its workers. Consequently, such work is seen as temporary and has very high staff turnover. Even in more general farming, long hours and tiring manual work can be a significant part of the job. Automation has massively reduced manual labor and farming is no longer a primary occupation, but there are still areas of hard labor. Research is very active, and many new developments are being rolled out. These include robots for thinning, weeding and spraying crops, visual inspection of crop health by drone, robot harvesting of soft and delicate fruits, as well as animal care (like the robot milking parlors mentioned in chapter 2). With better sensing systems, it becomes possible to respond differentially to individual plants and animals, thus increasing yield and productivity. While large, mass-production machines like combined harvesters are economic for volume processing, new robotic techniques are focusing on the more marginal and specialized areas.

Other areas of certainty can be seen as extensions of existing trends and developments. We will all be communicating with machines more freely through speech and images. Bio-identifiers, such as facial recognition, will speed up social interactions and communication. Shopping will become more (if not totally) cashless, with novel delivery of goods and remote ordering; automatic teller machines (ATM) will become rare. Personal devices, centered on the smartphone, will support, manage, and integrate more functions involving fitness, health, sports, entertainment, life management, family and friends. As these become integrated, it could become difficult to untangle individual services and switch to new suppliers.

New products will be easier and faster to manufacture because most of the processes have become digitally managed. This includes highly flexible production equipment like three-dimensional (3D) printing, robot warehouse handling, building maintenance devices, flexible assembly robots that are safe for human interaction, and all manner of test, packaging, and delivery systems. In addition, recent advances in AI offer software control and monitoring tools of unprecedented flexibility and management transparency. Cooperative robots (known as *cobots*) will find increasing use in manufacturing, as a result of the growing interest in human-robot interaction.

The cheap availability of deep cloud services will spread these methods far and wide. For applications that don't have safety-critical issues, we can expect the merging of speech, text, and image processing inside deep systems to provide agents that can talk to us, understand images, read text, and generally be useful assistants. So long as we can tolerate the occasional error, deep systems are likely to grow into powerful new applications. This is already seen in cloud-based phone apps. It is certain that Deep Learning systems will win a number of future races against existing AI methods.

TRUST, VALIDATION, AND SAFETY

Much of the fuss about AI comes down to a matter of trust. We want to trust our plumbers, accountants, doctors, and lawyers because trust is a powerful way of simplifying relationships and carrying out transactions. The amount of effort spent in checking out the background and reading all the rules and conditions for a new deal with an untrusted agency is enormous. It is even more complicated than this because just being labeled as a member of a trusted group doesn't necessarily mean that an individual can be trusted as much as the group. So if you want to find an accountant you could simply find anyone described as a "chartered accountant," but you would most likely do better to obtain recommendations from (trusted) friends who have experience of some accountants.

Our automated systems need to be trustworthy too, especially those that make important decisions for us. Unfortunately, trust is not a one-dimensional measure; it varies according to context. For example, in rock climbing, there is a trust issue known as "the person on the end of the rope." You might have a very dear friend who you trust implicitly with most things, but you just wouldn't allow him to be your partner on the end of the climbing rope. He might be an excellent accountant whom you trust with your money, but he is accident prone when it comes to physical things. Another friend might have a chaotic lifestyle and be hopeless at managing most things, but you know her to be an expert rock climber who knows how to deal with knots, belays, and safety. So your first friend can be trusted with your life savings but not saving your life, and vice versa for the second friend. Trust depends not just upon selecting the right agent, but also on what behavior is being trusted.

Experience teaches that trust has to be earned, or recommended through the experience of other trusted agents; it cannot simply be asserted, although the persuasive powers of the advertising industry try to do just that! It follows that new products must be validated in some way before they are released. This is where standards can play a vital role. All professional engineering branches have developed rigorous standards: imperative rules and requirements to ensure high-quality results and eliminate risk and damage.

For example, great care goes into car manufacturing; extensive failure analyses are carried out, exhaustive testing is performed, and the simulation of many design options ensures compliance with national and international standards. Multitudes of highly technical standards range from braking dynamics to driver visibility, from illumination specifications to noise levels. This creates high-quality products with high demand and economic value. And exactly the same is necessary for advanced AI and robotics.

THE PRODUCT-CENTERED VIEWPOINT

What does trust mean for AI and robotics? It means just the same as with most engineered products and systems; we trust them as useful tools if they do a good job, perform reliably, and don't cause any harm or damage. But this applies only to the (usually well defined) functions of the device or system. In banking, for example, we expect totally accurate and reliable transactions involving our accounts and various other basic banking functions. But it wouldn't be a good idea to assume the same level of trust about the return on a bank's market-based investment scheme.

A diagram can clarify the responsibilities and influences involved in trust. Figure 16.2 shows a company producing a product, together with an associated warranty. The customers rely on the product description, and their trust is reinforced by the terms of the warranty (shown as thick lines). The product is built to comply with appropriate standards, and these are compiled by a respected professional engineering association or similar body. To check that standards are met, independent regulatory agencies are often set up and supported by governments or international agreements (long-dashed lines). Products must not break any laws, and so various legal requirements are recognized in the warranties, built into the standards, and monitored by the regulators (these influences are shown as short-dashed lines). Ethical requirements, which may not have legal status but may nevertheless be decided by national agreement, can be incorporated by the same means. The customers also can have some influence on legal and ethical issues through consumer organizations or direct representation. Of course, standards are often designed to satisfy regulations, so a regulatory body is only needed as a monitoring process, but it

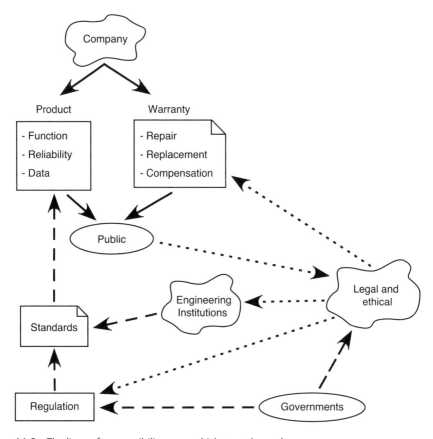

16.2 The lines of responsibility upon which trust depends.

is important that they have corrective powers so that any "substandard" products or malpractice can be stopped and treated appropriately.

The diagram indicates how legal and ethical concerns originate from governments, which often consult on opinions of experts, manufacturers, and public bodies before enacting new laws or regulations. This is a huge responsibility, as we all rely on these mechanisms to produce trust, without which life would be so much more difficult. Unfortunately, there are too many calls for debates and discussions between "technologists, policymakers, and stakeholders" that do not lead to useful results. Such meetings try to "initiate a dialog between industry, academia, and government to arrive at a mutually agreeable blueprint that quells concerns arising from privacy, security, safety, and ethical issues while not impeding innovation and technological progress."[15] While well meaning, these often miss the target. The problem is not to "quell concerns" while not stifling innovation; progress and innovation can't be suppressed that easily (research will carry on regardless)—and why should genuine concerns be quelled anyway?

Producing safe and ethical products is an engineering matter, not an innovative requirement. The only way that "privacy, security, safety, and ethical issues" can be satisfied is by building products and services so that they meet standards. Formal standards strengthen best practices, ensure high-quality performance, and constrain unsafe practices. It's really a matter of getting all the concerns (safety, ethical, and otherwise) properly resolved and defined in a standard that respects human needs.

From this product-centered viewpoint, a consumer considering a new product might ask: How can trust be evaluated? At a practical level, we can devise a few questions that help to expose the key issues.

Function—What Does It Do?
What is the purpose of a device or service, and what is it supposed to do well? Don't take any notice of how flashy it is, how well it chats with you, or how it will impress your friends; these are the issues that sales personnel love to tout. You need to find out exactly what function it can perform and what are the exceptions, boundaries, and limits. We don't need to get into all the complexity and fine detail; we just need a good appreciation (a mental model) of what the thing does.

Value and Risk—How Far Can It Be Trusted?
When you know what the device/application can actually do, you can assess how much use it will be for you personally, as well as how much trouble it could cause. This is the process of weighing the benefits against the costs. The costs are not simply monetary, and the risks are not simply reliability. The risk of failure is often known and can be used to calculate replacement rates, but having trust in the supplier means that it will keep up good standards of service and replace faulty items and repair damage caused by its products. If such trust doesn't exist (from warranties and conditions), then the costs increase to compensate.

Control—Who Is in Charge?

This is the question of who controls the system, the data, or both. Do you, as owner, actually have full control over the device/service? If it produces or collects data, where does it go, who owns it, and what access and rights to it do you have? It should be made clear who or what has access to any control decisions, operational information, and user data. So long as users have full information, they can decide responsibly about their use of technology and act accordingly.

These questions highlight the importance of transparency. Users have every right to expect that all relevant questions will be answered and they will be properly informed about the product or service. Unfortunately, there is a tendency to err on the side of secrecy. Companies cite "commercial confidentiality" and claim that they must protect their market interests. But actually, the free flow of information facilitates innovation and encourages activity and commerce. On reflection, it is clear that more transparency would reduce confusion and uncertainty and ease many current problems, conflicts, and corruption around the world.

Indeed, transparency may be the most important guard against a dystopic technological future. All dictatorships, criminals, and exploitative, divisive, and unjust systems of control depend upon secrecy. It is vital for these malign applications and abusers that the control mechanisms are not generally understood. Similarly, any controlling elite will have its own closed systems and hidden data: knowledge is power. Unfortunately, governments also have a tendency to think that secrecy is a necessary management tool, although some are more enlightened than others.

We have a great responsibility to use our skills well, and we should care about our human nature, our societies, and the wider world. And indeed, this more holistic and inclusive approach has proved to be more profitable and long-lasting in most cases. As every engineer knows, when designing a product, an operation, or a business, it is never possible to solve a problem by optimizing just one variable; money is only one factor among many that must be considered. An optimum result is obtained by reaching a justified balance between a set of desired outcomes, including both technical and human requirements.

We can try out our test questions on two really big areas for the future: the cloud and the Internet of Things (IoT). The cloud is conceptualized as a service that supplies digital storage and computing resources from a remote, global network with massive computing power. For businesses, homes, and smartphones, it is desirable not to have to run one's own computers. Taken to its logical conclusion, all storage and nearly all computing could be done on the cloud, thus giving users flexibility and mobility through portable, low-power access devices that serve up enormous resources. This is truly mobile computing: users could travel anywhere and work on the most powerful computing system using only a laptop or phone. The downside of the cloud is that it's a case of all eggs in one basket. It's not that hardware crashes would be a problem (the distributed network of servers covers that), but rather security and privacy. If you have some valuable private data, how can you be assured that it's safe on the cloud when you have no idea where it is, how many copies exist, and who else can see it? Clearly, general assurances aren't enough to engender trust, and detailed information on the control question is required.

The IoT is simply the idea that any and all appliances, devices, tools, systems, buildings and vehicles could or should be fitted with internet connections and be able to communicate directly with each other. The advance over what we have now is that the internet will be connected not just to people and computing services, but to every little switch, light bulb, gadget, and device imaginable. The manufacturers are racing to supply these interfaced products, seeing big profits ahead, and the number of things connected to the internet is estimated to reach 50 billion by 2020.

Smart homes seem to offer time-saving, convenient enhancements for everyday living. And they do give real help when we are sick or disabled, when someone is in the house using the facilities. But technology offers what technology can do, not necessarily what people most want or need. So the IoT becomes the vehicle for controlling home devices from near and far. Perhaps remote heating controls are very useful, but is an internet-enabled front door lock really a good idea? The most important function of a door lock is security, and one thing the internet is not is secure.

The claimed benefits for this instrumented universe include "new capabilities and services far beyond today's offerings, which will radically change how people go about their lives." This should have "major economic, social and environmental impacts; the intersection of which forms future sustainable growth."[16] But these sound like sales slogans, even though they are statements from academics. The small print in the IoT research agenda includes topics like reliability, security and privacy, disaster recovery, and self-organizing IoT. It's another case of the engineering roaring ahead of the basic science, which is desperately trying to catch up. The overriding concern with both the cloud and IoT, as well as many other new advances, is the same: security. We all want to be able to trust our technology, but security problems are trust-killers. From a product-engineering viewpoint, we must build in standards and regulation right at the beginning, and security is the number one feature.

THE CRUCIAL ROLE OF HUMANS

A very important but often underestimated aspect of computing is the fact that many of the most valuable applications or services do not just spew out machine-generated answers; rather, they are the result of complex and subtle interactions among humans and machines. It has long been known that humans plus machines are more productive than either alone. This is now being recognized for AI; for example, Daugherty and Wilson (2018) argue that AI will replace some jobs, but they believe the real power of AI will be in complementing and augmenting human capabilities.

This collaboration is not new, as any examination of modern manufacturing will show. When they build cars, planes, and phones, engineers use software tools such as computer-aided design (CAD) to produce, refine, visualize, test, and analyze complex designs of physical components (this includes electronics and the fabrication of silicon chips). But they do not simply write a specification, enter it into a computer, and receive a perfect design. The design process itself brings in constraints and problems that need human skills to resolve. Cars, planes, and phones cannot be produced by computers or humans alone.

Modern CAD systems are integrated with AI to provide sophisticated processes such as deep analysis of fault scenarios and creative design by examining ranges of options.[17] Engineers spend time working on CAD systems in a truly symbiotic relationship. And this applies to more than design; similar software systems exist for computer-aided manufacturing, computer-aided testing, computer-aided production—computer-aided everything! Although we say that we use software as a tool, and of course we do (for tasks such as for calculating or searching), often we have to guide and help the computer in the task, and sometimes we receive similar help and guidance in reaching the solution.

We find this interaction and augmentation in human affairs too. When my doctor tells me that a medical procedure has a 90 percent chance of a successful outcome, a 7 percent risk of minor complications, and a 3 percent risk of serious problems, I don't mind if these estimates are produced by computational data crunching, but I do want to talk to a human doctor about it. In particular, I like to ask my doctor: "Considering all this information, would you undergo this procedure if you were in my position?" I personally find the answer to this question valuable; it is a human response to the issue, and that's what humans want and need.

This is sure to be one of the big areas for the future deployment of AI. An example of early work is the ambitious plans of IBM to extend the Watson system to a wide range of commercial and business applications. Guruduth Banavar is the chief science officer responsible for developing Watson in this initiative, which the company calls "Cognitive Computing." This approach is very much about combining human-computer capabilities rather than just automating applications. Watson will augment and enhance human expertise by learning from massive amounts of data, reasoning toward specific goals, and interacting naturally with people to perform a variety of tasks. The system will be customized for a particular domain and be able to answer questions conversationally to extract knowledge, discover insights, and evaluate options for difficult decisions. The aim is to provide commercial services for a range of sectors, from health care to financial. IBM hopes that Cognitive Computing will be to decision-making what search has been to information retrieval.

Guru Banavar builds this approach on three design principles: the purpose that the system is to achieve, transparency in terms of data sources and conflicting interests, and the economic opportunity in terms of skills and knowledge. Perhaps the most advanced application of Watson is in medicine; areas like radiology and pathology are already producing machine diagnostic analyses that are more accurate than those from skilled radiologists (Topol and Hill, 2012). Watson does not provide a final decision but acts as a consultant; the final decision must always be given by a doctor who is therefore responsible for the medical treatment.

This approach carries an important message: AI results may be impressive, but relying on total automation is nearly always a bad idea. Time and time again, the combination of human and machine has proved more effective, in terms of better and more reliable results, than machines alone. In many fields, especially medicine, this approach can also deal effectively with the issue of trust.

Details of more examples of AI in these symbiotic roles can be found in Daugherty and Wilson (2018).

THE ETHICAL VIEWPOINT

Ethics is less about products and more about the uses, or abuses, to which products can be put. As we saw in figure 16.2, ethical issues can be incorporated into standards and regulations, but the origins of ethical requirements are more complex than relatively obvious product needs like safety protection. Ethics requires discussion and agreed conventions and standards. This process is supported worldwide by committees, conferences, independent review bodies, and government task groups. For examples of important international bodies, see the IEEE Global Initiative on Ethics of Autonomous and Intelligent Systems[18] and the Future of Technology, Values, and Policy (a council of the World Economic Forum).[19]

In the UK, as part of the government's Industrial Strategy,[20] four grand challenge areas were defined: AI, clean energy, electric vehicles, and the aging society. An Office for Artificial Intelligence is responsible for overseeing implementation of the UK's AI strategy, and Demis Hassabis from Google DeepMind is an adviser. An industry-led AI Council advises government, promotes AI industry-to-industry cooperation, and boosts the understanding of AI in the business world. For robotics ethics, see a widely circulated paper by Boden et al. (2017).

It is encouraging that many of the largest companies seem concerned about the ethical issues relevant to AI systems. As mentioned before, IBM is involved in research in this area; for more information, see Banavar

(2016), which describes beneficial applications and best practices for ethical AI. Mustafa Suleyman, a cofounder of DeepMind, is profoundly interested in ethical AI,[21] and a new research unit, called DeepMind Ethics & Society, has been established.[22] Importantly, the AI community itself is working on research agendas for safe and beneficial AI (see Russell, Dewey, and Tegmark, 2015).

However, we must not rely on the creators of AI or the producer companies to use their own ethical criteria and set their own ethical standards. There will always be companies that attempt to make money through cut-rate products and dubious practices. We see this time and again when a fault or weakness that would cost money to correct is allowed to get through into a product. Recent large-scale examples include falsifying vehicle exhaust emission figures, bundling software to exclude competition, and compromising privacy through lax data collection.

These examples, and many others, all involve software. Unlike physical products, by its very nature, software is opaque to most people who use it, so users are at a considerable disadvantage. Nevertheless, software can be subject to the same kind of manufacturing controls and standards that we expect from the other things we use or hire. Transparency will be more of an effort for software, but complexity is no excuse, and users must be in a position to understand the technology and the issues. Transparency can prevent abuses, damage and harm.

If we know the options, know what happens to our data, and can see who is benefiting from our actions, then we are empowered to choose (and enabled to reject) unacceptably negligent or criminal behavior. Of course, the average user may not be skilled enough, or motivated enough, to probe the workings of a system, however user-friendly the documentation. But in an open framework, it only takes one interested individual in a population to find a problem and raise the alarm.

Transparency is the intrinsic model of the open-source software movement, where computer code is made freely available. The reliability of the software is enormously increased because it is used, examined, and improved by many software engineers and enthusiasts. All these experts find and correct errors and so when each new release is issued, it will be an improvement on the last. And although these software programs are free, they are licensed and released through controlled processes.[23]

Open-source software has become so high quality that it is used in government, industrial, and commercial systems. For instance, the International Space Station uses the Linux open-source operating system for major command and control functions.

Standards must be industrywide; in fact, they must be truly global for maximum effect. However, international cooperation is difficult to achieve. We can see this in the case of internet traffic. Despite the great benefits of the internet, it is shameful that extreme violence, racism, and pornography are widely available and that global drugs and organized crime now take advantage of dark internet services. The internet originated as an open, unmoderated system, so the onus is on any service provider to apply effective moderation of the material that they disseminate (at the source, not after the fact). As this often doesn't happen, we can only conclude that ethical policies are not taken seriously. It is not censorship to protect children from accidentally encountering such extreme and affecting material. Standards are increasingly being called for, but governments often seem loath to interfere in such so-called markets.

Some good news is that interest in ethical issues for AI is burgeoning. In 2018, the House of Lords Select Committee on Artificial Intelligence in the UK produced a report on the country's use of AI.[24] The report emphasized the economic value of AI and concluded that ethics are vital in this area. Five principles were proposed "that could become the basis for a shared ethical AI framework." The chairman of the committee, Lord Clement-Jones, said that, "it is essential that ethics take centre stage in AI's development and use."[25] This is a very clear message from a thorough and thoughtful report. The problems aren't technical (they rarely are); they are ethical. It is the use (or rather, the abuse) of technology that has to be managed.

LESSONS FROM OPAQUE AND UNREGULATED MARKETS

A salutary example of the devastating damage that unregulated software can cause occurred in the financial markets. The disastrous algorithms behind the global crash of 2007–2008 were not products designed by engineers based on sound science; rather, they were ideas produced by "quants" (quantitative analysts), who believed they had created sound techniques which in fact were inherently probabilistic, unstable and not well understood. It was profoundly unethical to treat unsound methods as though they were reliable and fit for their intended purpose, whether negligently or intentionally. The respected expert in quantitative finance, Paul Wilmott, has shown that financial models are "too theoretical, it's far too mathematical, and somewhere along the way common sense has got left behind" (Wilmott, 2010). The math involved in quantitative finance is fearsome, requiring at least degree-level knowledge, but this incident represents a woeful lack of understanding. No scientist could get respectable papers published in other fields using the same math, applied to the same poor level of assumptions, as seen in quantitative finance.

As far back as 2000, Wilmott had spotted the flaw in the mathematical models that were based on vague data (Wilmott, 2000). He warned that "a major rethink [of financial modeling] is desperately required if the world is to avoid a mathematician-led market meltdown." That disaster came to pass, and taxpayers and investors are still paying for the recovery costs for the global financial crash. The banks have now been shored up by taxpayer money, but Wilmott is currently complaining that the markets remain unstable. In 2018, the banks were again criticized for promoting large overdrafts and overly encouraging credit card debt; total debt is now up to precrash levels again.

This concern is discussed in Cathy O'Neil's book *Weapons of Math Destruction* (O'Neil, 2016). She explains why shifting from human decisions to algorithms can be so harmful. Decisions on loans, job offers, and justice should not be based solely on rule-based processes or solving mathematical problems. O'Neil identifies three features of these algorithms: they are important, secret and destructive. The algorithms that ultimately caused the financial crisis met all those criteria—they seriously affected large numbers of people, they were entirely opaque (hardly anyone understood them, even their creators), and they destroyed lives. Perhaps the key word here is *secret*; if the algorithms had been transparent, then they could have been analyzed properly; they would have had some peer review, and they certainly wouldn't have been so unfit for their purpose.

Understanding risks is probably the most important skill to have in modern life. Or, put another way, misunderstandings of risk are frequently responsible for disasters and misery. And the only way to reduce risk is through education, transparency, respect for ethical and human values, and government actions in introducing and maintaining regulation and standards. As a US presidential report on AI and its effects said: "Technology is not destiny—institutions and policies are critical."[26]

OBSERVATIONS

- Predicting the future for a short period is OK, but don't follow the curve too far. You can't tell what sort of curve you are on when you get started.
- Reasonable certainties for the near future include autonomous electric vehicles, advanced agricultural robots, cooperating robots in factories, and deep neural network applications everywhere. AI will automate many office tasks in which large volumes of documents have common characteristics and can be processed by learned patterns. This is made possible by the recent reductions in error rates, so that the number of rejected items is not overly onerous.
- The hardest problems are always those that have to cover all eventualities. In engineering, the last 20 percent of performance always takes 80 percent of the effort. Thus, achieving completely autonomous vehicles will take longer than the early successes suggest, and deep networks will require a lot more research before they can be trusted with serious or safety-critical decisions.
- Trust is the basis of commerce, finance, society, and civilization. Standards and regulations are necessary for the systematic maintenance of trust, and these ultimately fall under the responsibility of governments.
- Ethical issues covering human values have to be resolved by consensus and then incorporated into product standards and regulations.
- Transparency is the key to avoiding disasters, increasing understanding and trust, and sharing responsibility.
- A good way to understand new technology is by asking questions about function, risks, and who has control over what happens.
- Technology doesn't dictate outcomes. People don't care about technology; they just want to know what it will do for them. Many impressive

products get rejected upon the market launch, while inefficient oddities become serendipitously adopted (e.g., message texting).
- Automation involving human-machine cooperation is nearly always more effective than pure automation. Such methods are also better by far at incorporating human values into the process.
- AI has made significant progress on individual learning and thinking, and Big Data is providing access to vast stores of cultural human knowledge, but social intelligence involving human interaction is less well funded and needs more attention.
- Future threats from digital technology are real and important. But these threats are not just from AI and robotics, but also from the entire digital infrastructure that we build and use. The safe and positive management of these risks involves key issues of trust, standards, and regulation. Complexity cannot be an excuse for obfuscation, and a vital part of the solution is transparency, enabling the full involvement of all those affected (namely, the general public). This means that we must overcome system inscrutability, and full and clear information must be made available so that an informed public can make reasoned and reasonable decisions.

17

FUTUROLOGY AND SCIENCE FICTION

My advice to those who wish to learn the art of scientific prophecy is not to rely on abstract reason, but to decipher the secret language of Nature from Nature's documents, the facts of experience.
—Max Born, *Experiment and Theory in Physics*, 1943, 44

Predicting the future is fraught with difficulties. The last three chapters of this book have tried to shed some light on various likely developments in the short-to-medium term. These have all been based on projections of current research and market trends. Unfortunately, some futurologists seem to make a living from taking very scary and highly controversial predictions, often proposing astounding scientific developments verging on science fiction. They argue that most of modern science was science fiction a few years ago, but that doesn't mean that *all* science fiction ideas are plausible and will come to pass.

We should consider a few of these predictions, at least insofar as they relate to future technology. A review of expert opinion is helpful in gauging these projections, and in this chapter, I present results from various sources. I often criticize the neglect of human values—the almost wilful ignorance of the vital context of the human and natural world by some proponents of high technology. So it is appropriate to begin by reminding ourselves of the importance of our human nature.

ARE WE SMART ENOUGH TO KNOW HOW SMART ANIMALS ARE?

This section heading is the title of a book by the respected primatologist Frans de Waal (2016), a scientist who studied many animal species with the aim of understanding how they think and how intelligent they really are. He has shown that most human-specific characteristics, like toolmaking, empathy, communication, and cooperation, can be found in other animals. Maybe this is not surprising; after all, we have evolved from the ape lineage, so we should expect to find our traits represented to some extent in our ancestor species. In his experiments, de Waal has shown that monkeys are sensitive to inequality in the sharing of food and have some sense of fair play. He argues that although they do not have actual language skills, they effectively understand human communication by observing body language, vocal tone, and gestures.

Another authority on primate cognition is Michael Tomasello, a famous developmental and comparative psychologist. His work with both apes and young children has provided remarkable comparative insights into the subtle differences between the species (Tomasello and Herrmann, 2010). In experimental problems that require two individuals to act together to get a reward, both chimpanzees and children will work collaboratively to solve the problem, but children will learn to cooperate much more than chimpanzees can manage (Hamann, Warneken, Greenberg, and Tomasello, 2011). The difference between collaboration and cooperation is subtle. Young children are intrinsically motivated to cooperate, not only to detect inequalities, but to ensure fairness, help others, and see that others are helped.

It has become common to assume that, in terms of intelligence, humans sit supreme at the top of a pyramid, with the great apes just below, then other mammals, followed by birds, then fish, and continuing all the way down to insects. Modern science shows just how inappropriate this view is, and in chapter 14, we saw that even self-awareness is far from being a human-specific feature. We know that we share most of our DNA with our nearest animal species, and we also share many of their cognitive characteristics and modes of behavior. Even Charles Darwin denied any qualitative superiority. It seems that the things that give us our edge are not just using tools, but the ability to design them, and not just our sense of self, but our grasp of the past and idea of the future. These very few extra characteristics allow us to strive toward imagined goals and sustain the patience necessary to execute plans well into the future. Such apparently tiny additions to our basic competencies open up a huge range of possibilities, allowing us to develop sophisticated interests such as music, sports, art, and literature, all of which do not play much of a role in survival.

Consider the way that animals affect the world. Most animals live in cycles of predator-and-prey relationships, and their populations tend go up and down in cycles. They expand into areas of new territory and contract when climate and other external factors are detrimental. They live with nature, and they live in the landscape. Humankind is different; humans change the landscape. Partly because our social ability is so much more organized, we combine forces to build skyscrapers, create transportation systems on land, sea, and air, and invent concepts such as schools and hospitals. Thus, we produce all kinds of transformative effects on nature that other animals don't. This effect is probably unavoidable and usually irreversible—Alan Moorehead describes how the inhabitants of a remote Pacific island were first introduced to an iron axe: once they experienced the benefits of metal, there was no going back to the stone axe. This irreversibility of culture transfer in the eighteenth century is beautifully discussed in his book *The Fatal Impact* (Moorehead, 1966).

WHAT KIND OF WORLD DO WE LIVE IN?

We frequently hear the phrase *the digital world*, which implies that our main problems concern computing and communications technology. I believe this is a flawed and deeply damaging viewpoint. Yes, we do live in a "digital age," but our world is still the human world of life on the planet Earth. In this context, digital technology is very much an embedded feature of our current human life. Perhaps we should learn to say: "Computers now "live" in the human world."

You don't have to be a biologist to see that the digital scenario is a lesser, more constrained world than the biological world of life. Of course, computer systems and software have grown to be very complex and seemingly impenetrable to most of us. But it is a manufactured domain: We build it, we decide how to use it, and, yes, if we really want to, we can sit down and understand every bit of it, read the code, analyze problems, and make changes.[1]

The "facts of life" is the missing element, left out of so many predictions about robots becoming protohumans. As we saw in part I, robots are not alive and do not die. As we go about our modern lives, we may have little interest in evolution but we certainly do care about life and death. Robots may well come to "know" these facts of life about humans and be able to reason about them. But life will not relate to their own "selves"—their own particular and much more limited view of their environment and situation. This is important to remember when considering the claims of futurologists.

FUTUROLOGY, EXPERT OPINION, AND META-OPINIONS

Futurology is all about prediction, and at the Singularity Summit in 2012, Stuart Armstrong, now of the Future of Humanity Institute, reported on a study of over 250 predictions made about the date of the singularity (when computers would become omnipowerful).[2] He concluded that experts are no better at predicting these things than nonexperts, mainly because experts become overconfident and exhibit bias. He showed that the only pattern of any significance was that most predictions suggested the singularity would occur 15 to 20 years ahead of the time the prediction was made. In other words, those researchers who believed that superintelligence was possible needed enough time for them to work on it and make it happen. Armstrong concluded that we must factor in much more uncertainty; there simply isn't enough evidence to produce strong arguments for such definitive future events.

One way of exploring the future is to ask as many people as possible for their opinion. This gives better evidence than relying on a few celebrity futurologists. John Brockman has produced a book entitled *What to Think about Machines That Think* (Brockman, 2015). Brockman asked a large number of people to write 2 or 3 pages on their thoughts about thinking machines. He ended up with a 500-page book containing 186 short articles.

I classified the authors of the articles in two ways: the discipline from which they seem to be engaged or qualified, and their expertise in terms of three categories: research, management, or journalism. So an academic would be classified as someone with research skills, while a company director or institute leader would be a manager. The journalist category includes writers and authors. The results show that 70 percent of the contributors were researchers, 10 percent were managers, and 20 percent were journalists.

An analysis of the disciplines of expertise provides some interesting insights (see figure 17.1). One might think that the most prominent experts with opinions on machine thinking would be AI and computer people, psychologists, or philosophers—especially philosophers, as they are professional thinkers who also think about thinking. However, it turns out that less than 5 percent of the respondents were philosophers, 14 percent were computing and AI experts, and 24 percent were psychologists (this latter includes cognitive scientists and neuroscientists). Psychologists were the largest group overall, but even if they are combined with the AI/software people and the philosophers, they still account for less than half of the opinions. The next biggest group turned out to be

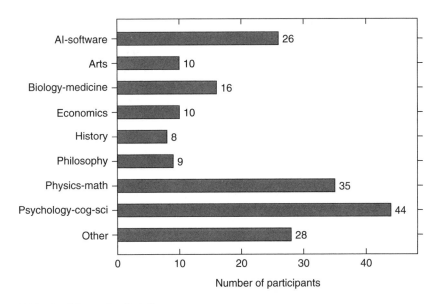

17.1 Participants by discipline.

mathematicians and physicists, including astronomers and cosmologists, at 19 percent. This is rather surprising because, although the physical sciences provides a strong foundation for reasoning about the world, they are not really qualified in either AI or the life sciences. We might expect that the biologists and medical experts would have quite a bit more to say about thinking, and the brain in particular, but they only accounted for 9 percent The remaining subjects (history, economics, and the arts), were all around 5 percent each, and there was a large group called "other," which mainly consisted of people who did not give enough information to assess their background. Interestingly, just over 20 percent were from Europe, with the remainder from the United States; and 15 percent of the total were female. These figures have a tolerance of a few percentage points because some authors did not give complete information.

I noted the recurring issues and recorded statistics on 15 facets of opinion, as listed in table 17.1.[3] These facets were scored across all contributors; either support or rejection of the issues were counted, and neutral views were ignored.

Table 17.1 Key for bar chart in figure 17.2.

Label	Issue
Progress	Future progress in advanced AI will be fast (slow)
Limits	Some serious limitations exist (no limits)
AGI	Probability of AGI emerging is likely (unlikely)
Super-AI	Probability of super-intelligence emerging is likely (unlikely)
Cyborg	Cyborgs/implants are important (unimportant)
Trans-h	Transhumanism supported (rejected)
Harm	Outlook for humanity is benign (malign)
Awareness	Self-awareness is important (unimportant)
Consciousness	Consciousness is important (unimportant)
Motivation	Motivation seen as a problem
Embodiment	Embodiment important (unimportant)
Values	The need to incorporate human values (not important)
Empathy	Importance of empathy in AI (unimportant)
Society	Importance of social networks, human-cooperation (unimportant)
Responsibility	Human responsibility and involvement is important (unimportant)

Note: Opinions were scored as either showing support or rejection for the issue.

FUTUROLOGY AND SCIENCE FICTION

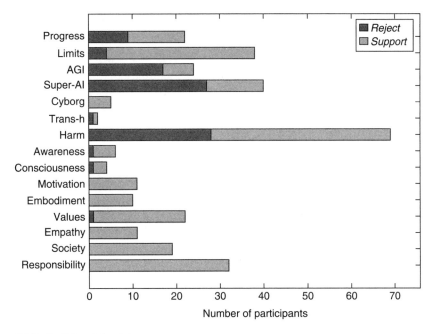

17.2 Participants' opinions.

Figure 17.2 shows the opinions, which have some interesting areas of consensus. Most articles only addressed one or two of the topics, so the total size of the bars indicates the popularity of the various topics. The first opinion concerned the rate of progress; there was an overall view that progress in AI will continue to be fast rather than slow. But the idea that there are limitations on producing thinking machines was very strongly advocated by a majority (89 percent). There were more opinions about superintelligence than there were about Artificial General Intelligence (AGI), but in both cases, there was a strong bias toward improbability. The biggest response concerned the question of harm—that is, whether the outlook for AI looks benign or damaging for humans. A total of 35 percent of the participants expressed an opinion on this point, with a big majority having a positive outlook. The importance of incorporating human values was very strongly supported with 95 percent recognizing this key ingredient. Human responsibility for the design and development of AI was unanimously supported, as was the importance of machine-human cooperation.

It is to be expected that a person's opinion is likely to be influenced by his or her background.[4] Looking at just the larger categories in figure 17.1, it was noticeable that different groups give different results. The biggest group, psychology, cognitive science, and neuroscience saw a predominantly benign outlook (76 percent), with no one expecting AGI. The software people and the biologists had similar benign views on outlook (60 percent and 63 percent, respectively) and little support for AGI. The group of physicists and mathematicians were split evenly on outlook, and only 28 percent saw AGI as probable, but 57 percent expected superintelligence.

It was interesting that within the physicists, there was a small group of cosmologists or astrophysicists, and these people were unanimously of the view that the outlook was bad and superintelligence is on the way. It seems reasonable that whatever field one has spent many years studying is likely to have some related influence on one's opinions, and thus correlate with others with the same career history. But why these particular groups show these particular predilections I leave for readers to ponder.

THREATS ON THE HORIZON?

It seems that expert opinion is leaning toward a benign outcome for future AI technology. In the previous section, we saw that most opinions were positive, with social, ethical, and human values clearly being stressed as important. This is encouraging, but we should also look at some of the various doomsday scenarios. Let's start with the most extreme, and then move to the more likely forms of dystopian outcome.

SUPERINTELLIGENCE AND THE SINGULARITY

Superintelligence has been proposed as a form of AI that is able to continuously self-improve. The idea of the singularity is that autonomous AI systems will become so powerful that they will be uncontrollable. Then, in a kind of intelligence chain reaction, they will make humans redundant and take over the world. This catastrophic end-result of current AI—nothing less than the displacement of humanity by an evolving, superior digital force—has been a viral meme in the world's media for some time.

I admit to difficulty in seeing superintelligence as a viable scientific possibility.[5] Consider three main problems:

1. **Superintelligence requires AGI as a prerequisite.** But this is nowhere on the radar. It really is necessary to see some results on this (even concrete designs would help) before claiming what it will do and what the consequential effects will be.
2. **What are the mechanisms for super-AI?** The idea of superintelligence has been well described, but the likely machinery behind it is hazy. In *Superintelligence*, Bostrom (2014) suggested several paths for reaching this state. These include AI methods and Whole Brain Emulation (WBE), both of which I've argued against in chapters 9 and 10. Other routes, such as biological modification and collective intelligence, are not very convincing because there is even less evidence for the possible processes involved. And evolutionary programming is much more limited than it sounds.[6]
3. **For the singularity, superintelligence will need to be autonomous; it must set its own goals and have its own agenda and needs and wants.** So how are motivational preferences for destructive goals to be provided? I've argued that selfish, malign behavior is not relevant for robotics because the full range of human emotional states are not appropriate, nor are they of any value. With a lack of competitiveness, endless patience, and a limited ability to harbor negative emotions it is hard to envisage how a plot to overthrow humanity might be hatched.[7]

It is difficult to see how such systems might attempt to appropriate a nation's electrical system, or take over the internet, without any opportunity for humans to interfere in the process and correct the situation. Most systems have human controls and points of intervention; even the power lines running across a country have manual circuit breakers. When humanity is under attack, it always defends itself, and history shows periods of great activity and innovation during such times of mobilization to meet major challenges. For example, we could disconnect the internet if absolutely necessary; we could survive without it and would find alternative (perhaps temporary) communication means. It would be (extremely) uncomfortable, but comfort is not as important as survival.

Chaos and disruption would certainly follow an internet failure, but that would be preferable to being totally subservient to some unfathomable digital superpower. The idea that we could not recover and defeat a rogue intelligent system shows a poor opinion of human nature. An authoritarian government might be able to impose such a system on a population, but a digital system alone wouldn't have a chance against hostile people.[8]

TRANSHUMANISM—DOWNLOADING THE BRAIN

It is sad that the twenty-first century has seen a general increase in antiscience and the denigration of expertise. Opinions, beliefs, and even falsehoods are sometimes promoted as facts when there is no supporting evidence at all. Thus, science fiction ideas can spread in popular culture as possible (or even likely) events. People say, "Well, anything is possible, isn't it? Why not? Who knows?" Well, actually, a lot of things aren't possible—electricity generation by perpetual motion machines, for example. And why not? Because we have thermodynamic theory, which explains very clearly why not. And who knows? Well, anyone and everyone who takes the time to learn (or remember) the specific bit of science that deals with that idea.

Scientists have toiled over these theories, experimented for years, and amassed piles of evidence. They know why some things are feasible now, while others will need a few years of work, and others are intractable and not worth starting. Expertise is not simply the opinion of a group of professionals—it is collective, carefully tested evidence that forms the very best knowledge we have at the present time.[9]

There are many topics in this category that impinge on robotics and AI, some of which originate from futurologists who should know better. There is only space to mention one example: transhumanism. This is founded on the assumption that everything that makes up a person—the individual self, the life history, the sum total of experience—is contained within the brain. The brain is an electrical system that can be encoded in digital form, and therefore it should be possible, or so the story goes, to download a complete copy of a brain into a digital computer, thus capturing the entire personality of that particular person. At some future point, this "digital version" of a person could be reenergized in some kind of simulation, and then the person would "live" again.

But let's consider this scenario logically. Suppose that a scanner could record the complete state of a brain at an instant in time. The captured digital data is a snapshot, a cumulative result of a cognitive life as it stands up to the point when the scanning was performed. But how is it to be used? A snapshot does not tell you what happens next.

Now suppose this digital "brain" is to be uploaded and restarted. As I've argued, humans and robots are so fundamentally different that they can never totally replace one another. So the only way transhumanism could work would be by uploading the digital brain into a human brain. This is the really difficult bit. The process would have to be very fast to set up the 20 billion cortical neurons into the exact state recorded in the downloaded snapshot. This requires some fantastic invasive technology that has not been proved even theoretically feasible. Also, the neural wiring diagram of one person is not the same as another, so how are these circuitry mismatches to be handled? All of this seems very remote; after all, even the scanner to get all this started is not yet realizable. And the process would be so unavoidably unpleasant for the recipient that hopefully it will never happen.

Like cryonics (the freezing of bodies in a kind of suspended animation), transhumanism is an attractive idea to those who wish to become semi-immortal. Such ideas are not new; George Bernard Shaw produced a play on this theme in 1921 (*Back to Methuselah*), but they have no scientific credibility. It should be clear by now why transhumanism simply won't work. Modern robotics has shown how important embodiment is. Without a sensory-rich body, perception as we know it is impossible (never mind sentience). And enaction is also vital; our actions are entangled with our thoughts just as much as our feelings. The life process, the life cycle of the individual, cannot be separated from embodied cognition. This is the difference between biological brains and computer brains.

IMMINENT THREATS

It can be harder for a journalist to cover [stories about] hope rather than fear.

—Laura Kuenssberg, BBC Radio 4, April 27, 2018

A really frightening aspect of AI technology is the damage that it could do when in the hands of malevolent humans. Just as the combination of humans and machines in close cooperation can, and has, delivered tremendous benefits and advances, so that same partnership has the potential to wreak awful suffering and destruction. Consider the algorithms already working away in the internet, collecting information for news sources and feeds to social networks and media sites. The criteria for this news selection should emphasize measures like truth, accuracy, integrity, as well as having priority, and relevance. But increasingly, we have seen populism driving the selection criteria, perhaps more than truth. News organizations can't seem to resist the pull of fashionable and popular issues, no matter how trivial, misleading, or offensive they may be. This is how "fake news" becomes disseminated and established, and how a single (erroneous) report can turn people away from lifesaving preventive medicine; the scare about the triple MMR vaccine (for measles, mumps, and rubella) is a prime example.

The main social media platforms have already been hit by complaints, court cases, and substantial legal fines for their poor management and control of personal data. They have been accused of "human hacking," otherwise known as *social engineering*. Although this term has recently been applied specifically to security crime (illegally extracting data from people), the original concept from the social sciences means any psychological technique that influences populations toward particular outcomes without their agreement or knowledge.[10] In the 1950s, psychologists discovered a phenomenon known as *subliminal perception*. They found that if an image was flashed up for a fraction of a second during a film, the conscious mind would not register it, but it would nonetheless be seen by the subconscious mind and affect future behavior.[11] This proved to be so insidious and effective that it has been banned or made illegal in most countries. This is an example of social engineering—controlling peoples' desires, wishes, and preferences by means of which they are completely unaware and don't have the option of avoiding.

But isn't this exactly what is happening today? Social media, trading companies, and other organizations involved in harvesting personal data are no less engaged in social engineering, but apparently the law either does not apply to them or has not yet caught up. For example, it may be quite acceptable to collect data on people's addresses and bank accounts when you're trading with them and transactions are actively taking place. But if this data is later used to influence their choice of products that they buy without their knowledge or agreement, then this is a form of subliminal persuasion. It's probably worse than we think—or want to think about!

We are all being spied on by these means: Our movements are being tracked, the places we visit are recorded, the products we buy are logged, and our preferences are being assessed by measuring the time that we linger over a particular product or event. We are being spied on without our knowledge or consent. What purpose could this data be used for? But if it can increase your profits, sway your political margins, or reduce crime and improve citizenship, isn't that a good thing? Why should we care? *The Age of Surveillance Capitalism* (Zuboff, 2019) offers some convincing arguments as to why we really should care.

The real danger in social engineering is that it can become a tool for the control of populations. China is experimenting with linking benefits and rewards to good behavior. Capitalist societies already have this in a weak form, with credit ratings controlling access to mortgages and so on, but the idea of reward and punishment connected to digitally gleaned metrics of good citizenship harks back to George Orwell's dire warnings of state control.

The only way around this problem is either some form of transparency, whereby users can see all their information everywhere it is held, or active regulations, which prohibit and penalize organizations that misuse their data. In either case, it's a question of secrecy. The lack of access, the need for transparency, is the important issue here. Obfuscation by any means is the key to criminal activity. Because these methods are insidiously creeping into the news and entertainment media, social media, and the advertising realm, we do need to look for better standards and principles that can be applied to our current digital, data-centric world.

Unfortunately, we cannot leave this to companies, who will cite "commercial confidentiality." Regulators are needed to enforce legislated standards. That means governments must be involved. Furthermore, global superregulation is necessary in order to deal with worldwide digital technology. Some countries may not sign up but, just as for other global issues, group pressure can help to encourage compliance.

Note that social engineering can also achieve positive goals. So long as influences on people's behavior are minor and easily avoidable, usually by providing free choice, then it is often possible to reinforce positive values in a society. This small version of social engineering, known as *nudge theory*, has had many successes in increasing tax returns, improving reply rates to requests for information, and other efficiency gains for governments and companies (Thaler and Sunstein, 2008). The key point is it is not secret, people can see it happening, and they still have choices.

Once again, we see that we must insist on truth and transparency—no "alternative facts," no "confidential" sources or processes. We have the human right to know, over and above any excuses of "digital complexity."

TOWARD DYSTOPIA?

Disruptive technologies are those that have influence well beyond the particular application area from which they originated. They tend to cause sea changes in areas like social or business behavior, or governmental or national policy.

AI-based office automation is predicted to cause wholesale loss of jobs and serious effects on national and global economies. But what would it really be like? Would mass unemployment be a nightmare, or would we all enjoy more leisure time?

In November 2015, the chief economist at the Bank of England, Andrew Haldane, stated that AI applications might take 15 million jobs away from human workers in the UK. That's nearly 50 percent of the total! Similar predictions have been made for the United States. These jobs are not in the usual industrial target areas, but mainly apply to the white-collar workers in administrative, clerical, and production positions. Accountants were identified as being particularly affected, with 95 percent of their tasks becoming automated. The routine processes in accountancy, law, and other professional areas could form the basis of automation, with the variations and customizations involved being handled by advanced AI methods. The more physical human services, such as hairdressing, restaurants, and car maintenance are much less at risk, however, as are specialized, knowledge-based roles and creative occupations.

A key to detecting a threat to a job is to ask: Does this job involve a large number of repeated, very similar activities, or is the same process applied to a series of very similar objects/data/people? Thus, agriculture is a target for automation because picking fruit involves the same process, applied to many similar objects, and any natural variation can be learned and accommodated. But services in people's homes (repair work, for example) incorporate too much variation and are unlikely to be automated.

There is a wide range of opinion about imminent automation. One view is that we will end up with machines doing all the work, leading to mass unemployment and a rich management elite who own nearly everything. At the other extreme, because all the boring work will be done by machines, humans will enjoy a range of interesting new jobs as yet unimagined (or unimaginable), and wealth will be shared out so that most people have a comfortable existence. The former case may appear more likely, given the essentially selfish financial behavior of global markets, but the latter has some historical support too.

For example, automatic teller machines (ATMs) were introduced by banks around 1970, and the data for the next 40 years shows that 400,000 bank teller jobs were removed (Bessen, 2016). However, there are still as many bank employees now as there were in 1970. This is because the skills have changed; tellers no longer just count out money, but also deal with customer relations and assist customers with more technical queries. The banks found that interpersonal relationships are important to their business, showing that staff roles can change instead of merely being eliminated.

Some economists, like Guy Standing (2016) and Paul Mason (2016), believe that new methods of distributing income will emerge as a result of the current surge in automation. Because jobs will be lost, many will suffer a loss of earned income, and there will not be enough in taxes raised by a country's workers to support the current models of Western democracies. Standing and Mason advocate an unconditional universal income system, whereby everyone would receive a standard regular payment. Among other things, this would help the unemployed and balance the inequalities at the lower end of the employment spectrum. There are political objections to "paying people who don't work," as some commentators complain; but evidence (from retirees, volunteers, and hobbyists) shows that people actually *enjoy* working if they can find a suitable occupation, so maybe this idea is worth exploring in detail.

Anthropological approaches look at societies at large and analyze patterns in groups and societies. Joseph Tainter's book *The Collapse of Complex Societies* uses complexity theory and other modern tools to examine various ancient civilizations and how they survive and decline (Tainter, 1988). He believes that social complexity is important and related to healthy governmental organizations. The point of collapse can be reached when problem-solving institutions fail and societies lose much of their complexity.

Peter Frase covers two heavens and two hells in his book *Four Futures: Life after Capitalism* (Frase, 2016). He sees technology as largely neutral and the outcomes as being decided by political forces. Frase starts from the seeds of dystopia we already have: a small elite controlling most of the wealth and power, and global warming, which is already affecting global resources but is largely ignored. In the first "positive" future, unlimited clean energy has been achieved and automation is so productive and so egalitarian that no one needs to work. Resources are managed, carbon is controlled, and global warming is stabilized. Frase's second "positive" future is reasonably egalitarian and has features like a universal income system, but the carbonless energy breakthroughs have not yet been achieved and consequently, much political effort is required to change consumption patterns. Then there are the "negative" futures. The third future is "rentism," where powerful elites have managed to gain ownership of automation and intellectual property so that users have to pay for everything. As there aren't enough jobs, consumption is low and poverty is rife. In the final future, "exterminism," the superwealthy elite control all the automation and live in perfect conditions, while the planet warms and the masses have no economic role and become superfluous.

There are many books describing variations of these scenarios of the future, such as the popular but scary *Homo Deus: A Brief History of*

Tomorrow by Yuval Noah Harari (2016). It is worth noting that unbounded AI does not receive all the blame for a trajectory toward dystopian futures. The role of humans in serious misuse and abuse of non-AI systems is already manifest in modern life.

IT'S NOT ALL DOOM AND GLOOM!

The late physician and statistician Hans Rosling became well known for presenting innovative documentaries on global health with titles like: *Don't Panic–The Truth about Population* and *The Joy of Stats*.[12] He used graphs and novel dynamic displays to show, with great enthusiasm, that the world, contrary to most opinion, is becoming a better place. Rosling used reliable statistics to show that global population growth is slowing down and looks like it is stabilizing. There are many fewer deaths from natural catastrophes, global health is improving dramatically in all countries, and altogether, the global standard of living is enormously improved over that of a century ago.[13] Rosling's argument is that we have a kind of instinctive pessimism; we notice the threats and the worries more than we do the good times. This is reinforced by the media, which struggle with uneventful, upward trends. The title of his posthumous book, written with his son and daughter-in-law, is appealing: *Factfulness: Ten Reasons We're Wrong about the World—and Why Things Are Better than You Think* (Rosling, Rönnlund, and Rosling, 2018).

Another source of optimism is found in the work of Steven Pinker. Pinker argues that violence and violent death are in decline and much less of a threat in modern life than in the past (Pinker, 2011). The basic point here is that tribal and casual violence reduces as national governments gain control and enforce laws. Slavery, torture, and tribal clashes are much reduced, and, in the ideal, only the state executes war or violence. This leads Pinker to suggest we should use our talents to enter a new age of enlightenment, where the combination of reason and human values drives a progressive and peaceful age (Pinker, 2018).

These encouraging findings are not without detractors, but they are relevant to current concerns and the decisions that affect us all and require our involvement.

THREATS IN PERSPECTIVE

In terms of time scales, superintelligence, if it occurs, is likely to emerge at the earliest by 2070 (at 50 percent probability) (Müller and Bostrom, 2016). But one of the more immediate and pressing threats is global warming; the rate of change for global warming indicators suggests that we will experience severe geoeconomic crises well before that date.[14] If sea levels rise by only 30 feet,[15] two-thirds of the world's cities will have severe problems, and many of us will have "a house by the sea."

Meanwhile non-AI technological changes are also happening very fast, and problems like cybersecurity, economic imbalance, nationalism, and extremism (all supported by digital technology) are so real and immediate that any threat of robots overtaking the human species seems rather distant and low in priority. As Andrew Ng, chief scientist at Baidu Research (China's huge AI and internet company), said about AI and robotics threats, they are "like worrying about overcrowding on Mars."[16]

FINAL REMARKS

All new and powerful technology has the potential for abuse, and this is a real and serious threat for the deployment of modern AI. However, there are encouraging signs of informed concern coming from the main players and their stakeholders. This includes the AI companies themselves (Google and Apple are the most notable in this respect), the leading AI software developers and engineers, and politicians. There is a new emphasis on "friendly AI"; that is, AI that has been designed to perform and behave in accord with human values and criteria. This is likely to influence AI software design for the better.

Work is under way on the ethics and regulation of new applications. Public discussions, companies, and various agencies are producing reports, and governments are holding inquiries that result in recommendations and directives. Institutes like the Alan Turing Institute[17] are organizations that have recently been formed to address these important issues.

It is good that the main focus is on ethics, rather than just economic growth and commercial interests. However, it is imperative that this work on ethical and human values continues, and that it produces effective standards and regulatory systems. These are global issues, and much international collaboration needs to be negotiated. It is vital for these efforts to succeed, and they need all our support—they are our means for avoiding the AI dystopia about which we have heard so many warnings. We live in a human world, not a digital world, and we must assert our right to organize digital technology according to our best interests, not the other way around.

Disembodied AI agents will become very common in applications and provide powerful services that replace some existing human roles. They will talk to us and learn about and remember our personalities. They may pretend to be human (or sometimes be mistaken for human), and they will be accepted as useful assistants. But without a body, the full engagement of conversation will be lacking—social interaction cannot be synthesized from data, and an embodied model of self is necessary to fully relate to humans.

On the other hand, no embodied robot will ever be mistaken for a human. Even if constructed with the most authentic appearance, as seen in some of the remarkable Japanese robots mentioned in chapter 3, the dynamics of movement and behavior produced by engineered machines will always have many levels of mismatch with biological systems. This is not a problem for most robot applications. For more humanlike intelligence, like social interactions, a body-model and an understanding of self become a necessity in order to make sense of the human condition and interpret human agents. This is where the differences between robotics and AI become significant. Social interactive intelligence needs much more research, but part II showed how real progress can be made in this area.

While new AI carries threats, existing digital systems are also in need of urgent attention (e.g., cybersecurity, data integrity, mass manipulation). In addition, there are competing alarm calls from nondigital global threats that are even more worrying and pressing. Threats to humanity must take the highest priority: If we do not focus on the survival of the human world, then no one will care what happens to the digital world.

From our varied considerations about the threats and opportunities that advanced AI and similar technologies hold for humanity, it seems that transparency really is the most important factor. From this flows public awareness, a better understanding of complex engineering, and open systems that promote consensus standards, high reliability, and trust. It also makes criminal or accidentally bad outcomes less likely. We need high integrity in all our systems (both computer systems and human systems), and transparency leads to best practices and the elimination of errors.

Ben Shneiderman is a respected researcher into human-computer interaction who has made major contributions to the design of user interfaces. In his book *Leonardo's Laptop*, he probes the boundaries of interaction and possible new developments by asking how the genius of Leonardo da Vinci would react to modern digital opportunities (Shneiderman, 2003). He stresses the importance of public awareness: "As users, we must be alert to the dangers in our lives, make our best effort to stop malevolent applications, and support processes that limit their growth. As developers, we can choose our applications carefully to show positive contributions, open our thinking to anticipate unintended negative side effects, and expose our work to participatory design reviews. Open discussion, open code, and open ideas are among the greatest benefits of information and communication technologies" (Shneiderman, 2003, 243).

More recently, James Bridle made a plea for openness as the way to manage technology for our benefit, not our demise: "All of them [the current economic, political, global, and environmental crises] are the product of a general inability to perceive the wider, networked effects of individual and corporate actions accelerated by opaque, technologically augmented complexity…Our understanding of those systems, and of the conscious choices we make in their design, remain entirely within our capabilities. We are not powerless, not without agency. We only have to think, and think again, and keep thinking."[18]

We must insist on transparency. We must be allowed to understand modern systems, to accept our responsibilities, and to freely express our opinions. We must assert our humanity and our human needs. And we must keep thinking.

Appendix

PRINCIPLES FOR THE DEVELOPMENTAL APPROACH

Cognition depends upon the kinds of experiences that come from having a body with particular perceptual and motor capabilities that are inseparably linked and that together form the matrix within which reasoning, memory, emotion, language, and all other aspects of mental life are embedded.
—Esther Thelen (2000)

Part II of this book described a methodology and some experiments on developmental robotics. Such controlled laboratory experiments, both successful and less so, produce a range of results that are analyzed and interpreted to give insight and understanding. It may be useful for some readers, who might wish to become further involved, to bring this experience together in terms of some succinct developmental principles. Actually, not many of these are fully formed principles, but they capture and illustrate the main concepts, offer advice, and provide technical insights from hard-won research experience. Hopefully, these will be relevant for those interested in the design and building of developing robots, and even in starting their own implementation project.

The key requirements are embodiment, constructive cognitive growth, and enactive behavior.

Interactions, both physical and social, are vital for cognitive growth and shaping behavior, so bodies are essential as the substrate for perception and interaction. Mental growth requires general cognitive structures with maximum flexibility in order to support multiple functions. Action, driven by intrinsic motivation, is the force behind development.

Keep a holistic view.

Development involves the whole agent; all subsystems need to interact and work together and will be involved in all learning. This means mechanisms should be as general as possible and apply to all levels and systems.

Forget about designing robots for specific tasks.

Don't think of robot design in terms of tasks and goals. Humanlike robots will be driven by internal motivation and have a general, open-ended, and task-free learning approach. This means that designers should mainly be concerned with motivational mechanisms and general learning methods for dealing with novel situations and problems.

Employ simple models—compact and transparent representations.

It is important that the models used are easy to analyze and relate to the results produced. This requires transparency and low complexity. Aim for the minimal and simplest set of initial principles or mechanisms and build complexity only as needed to support the current level of interaction. If you think you need to solve two problems, is there a solution to another problem that might cover both?

Play behavior is essential for the growth of skill and intelligence.

Play can be seen as the main intrinsic driver for developing robots. It can be generated through the probabilistic variation of experience applied to new situations. This is a general algorithm that covers a wide range of behaviors, from early motor babbling, through solitary physical activity, to conversing and other interactive and social activities.

Play is not complicated.

If play is seen as the modulation of prior experience to fit a current situation, then it can be implemented by remembering action schemas that match parts of the current focus of attention and trying them out with elements that are either modified or replaced by features from the current activity.

Stages are qualitatively distinct, and transitions are achieved by overcoming constraints.

Timelines and schedules give important information about when and where constraints apply. The cognitive structures will grow and change with increasing stages, but the underlying mechanism of saturation and constraint lifting is a consistent process.

The contingency rule is a powerful learning tool.

If two events occur at the same time and this contingency can be repeated, then they are very probably connected. This criterion is a major learning rule.

Start at the beginning—you can't short-circuit development.

Development builds stage by stage, so make sure that you are beginning at the most appropriate point. In other words, any preprepared behaviors or intrinsic abilities must be fully justified by careful analysis of the developmental timeline.

Start small—then grow in resolution, refinement, and complexity.

Humans begin with very primitive perceptions and actions and then refine and combine them to create more complex structures. It is more important to get the general shape of the landscape before filling in the fine details.

Long-term memory needs to associate previous experience with current stimuli and actions.

An associative memory can operate in various ways: It can retrieve candidate actions to apply them in possible present and future contexts; it can suggest actions that might lead to desired results; and it can predict the outcome of a proposed action.

Intrinsic motivation can be implemented as novelty and curiosity.

A basic novelty detector makes for a very effective intrinsic motivator. This can be any event, action, or stimulus that is unfamiliar. Related methods for curiosity, unexpectedness, and surprise are closely related and give similar results.

Overlapping fields have many advantages over arrays of contiguous elements.

Such structures support superresolution, wherein much higher accuracy is obtained than that given by the spacing resolution of the sensing elements or pixels. They also provide compact representations of complex spatial transforms. And, most important, mappings between such arrays can be established with far fewer connections than are required with arrays of contiguous pixels. This means that both learning and using mappings can be achieved very quickly.

Be careful when using reward schemes.

Rewards can become a means of covertly setting up external goals. The play process will create goals through discovery and interaction (and scaffolding). Reward in these goal situations works well, but beware of building in reward sensitivities with a view toward accomplishing future goals.

Don't inadvertently build in any programming interfaces.

The only means of "programming" should be through interactive experience. Structuring the environment to select outcomes and shape available experiences (scaffolding) is likely to be the main way to train a developing robot to achieve goals.

Beware of any biological incompatibility.

Any method, however neat or efficient, that violates neuroscience or psychological knowledge is going to cause trouble later. This rules out many artificial neural network (ANN) techniques, extensive training, and all offline training. (Note that infants don't have enough skill or awake time to repeat anything hundreds of times.)

Simulation and simulated bodies are only useful for exploring ideas.

The unexpected, serendipitous, and detailed complexity of real life cannot be captured well enough in simulations to replace the real world. Any lack of authenticity in modeling the richness and interactive subtleties of actual bodily systems leads to the buildup of errors during the accumulation of experience and thought. Simulation is never 100 percent authentic, and many real-life events can only be experienced properly in the real, physical world.

Developmental robotics offers a way toward the social robot ideal.

The developmental approach, involving subjective experience and modeling of the self and others, offers an ideal test bed for building and understanding social robots. There is much ongoing research on robot/human interactions and social robotics, but this work, like AI, is fragmented across topics. Longitudinal experiments can provide the integration needed for developmental progress.

The internal correlates of robot behavior are directly observable.

In robotics, the experimenter is able to observe and record internal changes in concert with the external behavior. This "god's-eye view" allows access to the subjective experiences being built up and their correlation with the experimenter's objective perspective. This will also increase our understanding of social agents' models of the self, thus enabling this intriguing area to be explored scientifically.

Developmental robotics offers modeling opportunities for psychology.

It becomes possible to test hypothetical mechanisms for generating human behavior and compare them against experimental data, thus offering a kind of "computational model evaluation" for psychology. Shultz (2003) devotes a chapter to this idea in his book *Computational Developmental Psychology*.

Neuroscience provides a rich source of inspiration.

We gained very valuable ideas from the overlapping receptive fields that form cortical maps, and used similar overlapping structures as a learning substrate. Other inspiration comes from the sensory structure of the eye system, the attention mechanisms in the superior colliculus, and the structure of the tactile-sensing system, among other elements.

Developmental psychology is also a rich source of inspiration.

Brain science has been a major source of inspiration for AI in the past, while psychology has been much less popular. However, the developmental approach relies heavily on the vast body of data and theory on human behavior from experimental and developmental psychology. Also, developmental robotics can help to integrate developmental psychology and neuroscience.

Don't expect to solve all the key problems quickly.

Experiments in development are exploratory. It is a process of synthesis—exploring and refining postulated models and putative mechanisms. There is much work to do, and many opportunities and advances to be made.

NOTES

CHAPTER 1

1. Keepon was created by Hideki Kozima for the purpose of researching children's social development. The sensors (cameras and microphone) allow Keepon to detect human activity and respond in interactive scenarios. The head movement provides gaze control for attention and eye contact, while the body movements indicate emotional states—rocking from side to side (pleasure) and bobbing up and down (excitement). Kozima's research explores various stages and problems in children's social care and communication. See Kozima, Michalowski, and Nakagawa (2009) for details of this interesting work.

2. See King (2002) or https://www.youtube.com/watch?v=kie96iRTq5M.

3. See Smith (2016) or https://en.wikipedia.org/wiki/Silver_Swan_(automaton).

4. Note this comment from an expert on the complexity of motor control: "If we consider the skills of dexterously manipulating a chess piece it is clear that if we pit the best robots of today against the skills of a young child, there is no competition. The child exhibits manipulation skills that outstrip any robot" (Wolpert 2007, 511).

5. As an aside on the relation between the arts and the sciences, this topic stimulates reflections on the universality of beauty and truth. Anyone who thinks that science and engineering are not very creative fields of study could do worse than to look at some of the classical algorithms in computing. These are often very short, taking only a dozen or so lines, and in their simplicity and ingenuity, they are often very beautiful. Searching and sorting algorithms often contain no math and are good examples of such artistic invention. Algorithms are similar to logical or mathematical proofs; in just a few lines of text, a process or mechanism is captured, which may have amazing implications. Those artists who flaunt their ignorance of science as a

badge of their creative qualifications should consider the sheer creativity inherent in some of these structures.

6. Historical roots often contain the essential ideas. The reader may be surprised at how many modern computing techniques and systems were actually thought up more than fifty years ago. Apart from the fascination of such history, it is important to evaluate how problems have been tackled, sometimes abandoned, and eventually solved.

CHAPTER 2

1. The distinction being made here is narrow. A robot is autonomous if humans are just observing, perhaps through the robot's eyes—but any input of guidance or control means that it is not autonomous.

2. Husqvarna (https://www.husqvarna.com/uk/) is linked to the Flymo brand, which also make robot lawn mowers in the UK.

3. This is just a sample; there are many more: Lawnbott (Italy and US); Robomow (US); E.Zicom (France); AL-KO and Bosh (both Germany); Flymo and Husqvarna (Sweden).

4. They are improving too. (I bought too soon! I bought one of the very early wet-floor-cleaning robots from the US—it works very well except that its battery always run out before finishing).

5. https://www.which.co.uk/reviews/robot-vacuum-cleaners, accessed July 22 2019.

6. For an example of an industrial strength floor cleaner, see the CleanFix website: www.cleanfix.com/en/products/cleaning-machines/robo/ra-660-navi_p831.

7. For example, https://www.knightscope.com/ and https://yellrobot.com/robot-security-guards/.

8. http://endeavorrobotics.com/products#510-packbot.

9. Actually, the PackBot, like many other field robots, has a human operator who monitors and directs operations. We include these in the definition of robotics because most field robots have at least partial autonomy, and human involvement is necessary for monitoring the situation.

10. It must be quite an experience to see a 40-ton truck traveling at 60 miles per hour down the road with the driver reading a book; see https://www.daimler.com/innovation/autonomous-driving/freightliner-inspiration-truck.html.

11. An autonomous vehicle can also signal back to pedestrians and others with digital display boards (https:www.drive.ai/).

12. Google set up Google Auto LLC in 2011, and then in 2016 created Waymo for its self-drive work. Google's vehicles now have acquired several million miles of self-driving experience.

13. The Society of Automotive Engineers (SME) in the United States devised a standard definition of five levels of driving autonomy. See the J3016_201609 standard at https://www.sae.org/.

NOTES TO CHAPTER 3 321

14. Edge cases are used in engineering to test systems (and algorithms) at their limits, the idea being that if they work well at the extremes, then they should also work well in between the extremes. Tail cases can be different if each one has its own conditions, and not necessarily at any limit.

15. Kilobots are available from the supplier of a well-known research robot called the Khepera. Khepera are about the size of a large teacup and usually operate in swarms of up to 10, but the lower cost of the Kilobot allows swarms of up to 1,000 to be researched. For more details about Kilobots, see www.k-team.com/mobile-robotics-products/kilobot/introduction.

16. In 2007, D'Andrea was appointed as a professor at ETH Zurich, a top technical university in Switzerland. He then became chief technical advisor to Kiva.

17. Videos of the Amazon warehouse can be found at http://www.chonday.com/Videos/how-the-amazon-warehouse-works; and https://www.youtube.com/watch?v=cLVCGEmkJs0; and further details are at http://spectrum.ieee.org/robotics/robotics-software/three-engineers-hundreds-of-robots-one-warehouse. Alibaba, the world's largest retailer in China, also has a very similar warehouse (https://www.youtube.com/watch?v=FBl4Y55V2Z4).

18. For more details, see https://us.aibo.com, and for the history, see https://en.wikipedia.org/wiki/AIBO.

19. See www.sony-aibo.co.uk.

20. Unfortunately, Anki has ceased trading, but see a review of Cozmo at www.wired.com/2016/06/anki-cozmo-ai-robot-toy/.

21. https://www.toyota-europe.com/world-of-toyota/articles-news-events/introducing-kirobo-mini.

22. See https://www.jibo.com/ and https://www.heykuri.com/blog.

23. consequentialrobotics.com/.

24. *Robotics Today* is a robotics news outlet: www.roboticstoday.com/devices/all-a-to-z/a.

CHAPTER 3

1. For the evolution of this series of humanoid robots, see https://global.honda/innovation/robotics/robot-development-history.html.

2. https://global.honda/innovation/robotics/ASIMO.html.

3. See Gates (2007).

4. For Sony's history of involvement in robotics, see www.sony.net/SonyInfo/CorporateInfo/History/sonyhistory-j.html.

5. However, there are exceptions. Steve Grand, with the support of his family, accepted a precarious financial existence in order to build a homemade robot named Lucy (Grand, 2004).

6. Unsurprisingly, the marketplace usually adopts an "easiest-first" approach to developing new products, bringing out basic models followed by more advanced

versions. After all, if a company didn't produce an early basic model but waited until a better and more sophisticated product was ready, then a competitor might well jump in first and undermine the investment. This is a mistake sometimes made by academics who start up new companies.

7. The Robotics Challenge was held July 27–30, 2017 (https://spectrum.ieee.org/automaton/robotics/industrial-robots/aussies-win-amazon-robotics-challenge).

8. For recent innovation-stimulating events, see https://www.amazonrobotics.com/#/.

9. Dr. John Long leads the biorobotics work at Vassar College. See Long (2012).

10. Long-term shifts in the age mixture within populations are predetermined by current birth and death rates. This allows planning for school sizes and other services.

11. Indeed, very brief, spontaneous, and tiny facial changes, known as *microexpressions*, are associated with lying and deception, and these are now detectable by software that have the potential for use in security applications.

12. See http://spectrum.ieee.org/robotics/humanoids.

13. Honda showed a video of P3 walking down a busy city street, and the alarmed glances of the passersby as they tried to ignore it and remain unperturbed are quite amusing. For details, see http://world.honda.com/ASIMO/history/p1_p2_p3/index.html.

14. The name *ASIMO* comes from "Advanced Step in Innovative Mobility," which indicated the original research aims.

15. See https://spectrum.ieee.org/automaton/robotics/humanoids/iros-2017-honda-unveils-prototype-e2dr-disaster-response-robot.

16. See some extraordinary videos of dynamic stability at www.bostondynamics.com.

17. See https://ti.arc.nasa.gov/tech/asr/groups/planning-and-scheduling/remote-agent/.

18. See http://robonaut.jsc.nasa.gov/.

19. See http://spectrum.ieee.org/automaton/robotics/military-robots/nasa-jsc-unveils-valkyrie-drc-robot.

20. See https://secondhands.eu; Ocado (https://www.ocadotechnology.com/)is an online supermarket, and the target application area is Ocado's automated warehouses, which are similar to Amazon.

21. Rethink Robotics (https://www.rethinkrobotics.com), a German company, was cofounded by Rodney Brooks, an MIT robotics pioneer.

22. The scoring scale necessarily has to be crude because this is a subjective and qualitative exercise (however, a proper survey could be done with a carefully designed questionnaire and a large number of technical experts). Actually, a linear scale is too simplistic—a few points of difference at the lower end does not match the difficulty covered by a few points of difference at the upper end. But this is an illustrative

device, and it explains why some application areas have proven much easier than others.

23. For uncertainty and novelty, the most benign case is *negligible*; hence for these two features, we give *Negligible* a value of 1. Knowledge and order are best when we have *Maximum*, and so for these we set *Maximum* at 1.

24. People have been building robot ping-pong players for at least 30 years. There are a number of commercial training robots (that just fire balls at you), but the impressive FORPHEUS robot can play a full game. It uses cameras to track the ball and the human opponent and has a high-speed motion controller. This robot can assess the skill level of the opponent and play accordingly. See https://www.omron.com /innovation/forpheus_technology.html and www.guinnessworldrecords.com/news /2017/2/japan-tour-table-tennis-robot-earns-a-futuristic-record-title-463501.

25. It was unfortunate that the 2013 rescue robots generated some negative publicity when very expensive humanoids were seen to fall over while trying to open a door.

26. Hiroshi Ishiguro famously made a robot copy of himself (see https://spectrum .ieee.org / robotics / humanoids / hiroshi - ishiguro - the - man - who - made - a - copy - of -himself).

27. This is known as the *uncanny valley effect*; see Masahiro Mori's short but classic paper on this topic (Mori, 1970).

CHAPTER 4

1. The concept is quite old, being formally established in the 1950s, when the first electronic computers were being designed and built. It is now a very active research area involving the majority of the world's universities and a wide range of companies ranging from the very small to giants like Google.

2. In the 1940s, Turing worked on the idea of chess-playing machines, and, in 1953, he wrote about a method that is essentially the basis of chess programs today. See chapter 16 in Turing and Copeland (2004). Other computing pioneers wrote similar proposals.

3. Search trees are drawn upside down because the root node is usually the starting point.

4. If there are M moves available at every position, then there will be M^l positions generated at level l. Of course, the number of choices available varies at different stages of the game, but the averages are roughly similar for most games of chess.

5. For a list of past tournaments, see http://www.grappa.univ-lille3.fr/icga/game .php?id=1.

6. Deep Thought was developed by Murray Campbell, Joseph Hoane, and Feng-hsiung Hsu.

7. Players are often dissatisfied with the style of play produced by modern computer chess programs, and there have been some attempts to provide limited variation in playing styles. Some commercial programs can offer a range of playing styles, or "personalities" (e.g., Chessmaster; see https://en.wikipedia.org/wiki/Chessmaster). Not

surprisingly, these supposed "fixes" to the basic search technique have not produced humanlike play, not least because no one knows what mechanisms would even come close to effectively modeling human chess intelligence!

8. See http://www.2700chess.com/ for live ratings of the top players.

CHAPTER 5

1. Kahneman specializes in human decision making, particularly concerning choices, risks, and bias in economic and financial scenarios. His book is long, important, and very readable (Kahneman, 2012). Kahneman did most of his work with Amos Tversky, who died in 1996. Their remarkable collaboration is beautifully described in Lewis (2016).

2. Note that I use a subjective viewpoint for terms like *information* and *knowledge*—information has value to a recipient if it can be understood, whereas knowledge grows in individuals as they receive meaningful information. This is different from information theory, which is a physical-mathematical concept and finds information everywhere.

3. This is a complex area, but, briefly, Extensible Markup Language (XML) goes a bit further than HTML and describes, to some extent, the content of the message, and then the Resource Description Framework (RDF) is even more content oriented and tries to represent the inherent meaning. RDF has been described as a general-purpose language for representing information. We then get to Web Ontology Language (OWL) which is built using RDF and used to write ontologies.

4. This is bad for my laziness—my filing is atrocious because I'm no longer afraid of misfiling!

5. Amusingly, you can have a new mathematical theorem named after you. Edinburgh University, a long-standing center for AI and formal reasoning, offers the following service: "TheoryMine, one of the world's most advanced computerised theorem provers lets you name a personalised, newly discovered, mathematical theorems as a novelty gift." See more about TheoryMine at https://theorymine.co.uk/.

CHAPTER 6

1. The most studied part of the brain is the visual cortex. For more on visual processing in the brain, see Zeki (1993) and Miikkulainen, Bednar, Choe, and Sirosh (2006).

2. See http://image-net.org/challenges/LSVRC.

3. The basics of ANNs are covered later, in chapter 9.

4. For a short review, see LeCun, Bengio, and Hinton (2015).

5. See Mnih et al. (2015) for a starting point for more technical details.

6. For a detailed report on the event, see https://www.scientificamerican.com/article/how-the-computer-beat-the-go-master/ or DeepMind's website, https://deepmind.com/research/alphago/.

7. The automated reading of checks becomes economically interesting for a bank when the rejection rate rises above 50 percent.

CHAPTER 7

1. Extract from "Learning" entry in https://en.wikipedia.org/wiki/Glossary_of_education_terms_(G-L), accessed July 26 2019.

2. ML factored the subject into broad classes of learning, such as learning by analogy, learning from explanations, learning from examples, and learning through discovery.

3. These quotations are from Tesco's website (www.tesco.com). Financial data is from Tesco plc, Five year summary, https://www.tescoplc.com/investors/reports-results-and-presentations/.

4. See Tesco's website.

5. Extrapolated figures using articles like www.independent.co.uk/environment/global-warming-data-centres-to-consume-three-times-as-much-energy-in-next-decade-experts-warn-a6830086.html and Markoff and Markoff (2005).

6. See Demis Hassabis and David Silver, https://deepmind.com/blog/alphago-zero-learning-scratch/.

CHAPTER 8

1. Bobby Fischer springs to mind. Sometimes known as the greatest chess player of all time, there are more than 30 books on Fischer's chess career and turbulent life.

2. *The Hitchhiker's Guide to the Galaxy* was a BBC radio series (1978) written by Douglas Adams. A series of books followed, then a TV version in 1981, and a film in 2005.

3. *2001: A Space Odyssey* was produced and directed by Stanley Kubrick. Arthur C. Clarke was also involved and simultaneously wrote the novel.

4. HAL was designated as male. Another Series 9000 machine, SAL (appearing in the 1982 novel by Arthur C. Clarke, *2010: Odyssey Two*), was female.

5. This story continues with a recent 50-year celebration consisting of an issue of a book, *The Making of Stanley Kubrick's 2001: A Space Odyssey*, and a limited special edition of 1,500 copies consisting of four volumes of film stills, facsimiles of original production notes, the screenplay, and other features (Bizony, 2015).

6. Jack Copeland's book (Turing and Copeland, 2004) provides many of Turing's original papers, as well as very useful interpretations. It is also fascinating to read the original manuscripts, with their (now) unusual style, corrections, margin comments, and annotations. The extensive archive is at http://www.turingarchive.org.

7. www.reading.ac.uk/news-and-events/releases/PR583836.aspx see also http://www.aisb.org.uk/events/loebner-prize.

8. http://www.theguardian.com/technology/2014/jun/09/scientists-disagree-over-whether-turing-test-has-been-passed.

9. https://www.chatbots.org/.

10. An example of minimum chaos might be a mechanical process like arithmetic, where the input digits have minimal uncertainty, there are no novel events, knowledge of the multiplication task is very high, and there is no disorder in the processing environment.

11. These examples are taken from schemas 107 and 136 from the collection of Winograd schemas, at https://cs.nyu.edu/faculty/davise/papers/WinogradSchemas/WSCollection.html.

12. Winograd schemas are often tested by searching the internet for previous usage. Usually, the results give as many hits either way—or none at all!

13. For examples of AI systems getting basic common sense wrong, see https://docs.google.com/spreadsheets/u/1/d/e/2PACX-1vRPiprOaC3HsCf5Tuum8bRfzYUiKLRqJmbOoC-32JorNdfyTiRRsR7Ea5eWtvsWzuxo8bjOxCG84dAg/pubhtml.

14. Trigger was a road sweeper in a UK comedy series called *Only Fools and Horses* on BBC-1 TV. Trigger's impressive claim that he's still got his original broom after 20 years is later debunked when he adds that the broom has had 14 new handles and 17 new heads.

CHAPTER 9

1. Andre Geim and Konstantin Novoselov, at the University of Manchester, won the Nobel Prize in Physics in 2010 "for groundbreaking experiments regarding the two-dimensional material graphene." (For more information, see https://www.nobelprize.org/prizes/physics/2010/summary/.) Graphene is a single-atom-thick, laminar form of carbon. Geim and Novoselov pulled graphene layers from graphite and transferred them onto a silicon wafer by a process known as the *Scotch tape technique*. Interestingly, another form of carbon had been discovered earlier, for which Harold Kroto, Robert Curl Jr., and Richard Smalley, shared the 1996 Nobel Prize in Chemistry. This form of carbon was in the shape of a geodesic sphere, as invented by Buckminster Fuller, and was called *buckminsterfullerene*, *buckyballs* or just *fullerene*. Fullerene didn't live up to the media hype expectations; let's hope that graphene does better.

2. See the EU website for project details, https://www.humanbrainproject.eu/en/.

3. See this article in *Scientific American* for more background: www.scientificamerican.com/article/why-the-human-brain-project-went-wrong-and-how-to-fix-it/.

4. See https://www.humanbrainproject.eu/en/brain-simulation/.

5. D. Deutsch, *Creative Blocks: The Very Laws of Physics Imply That Artificial Intelligence Must be Possible. What?s Holding Us Up?* Aeon, October 3, 2012, available from https://aeon.co/essays/how-close-are-we-to-creating-artificial-intelligence.

6. Other work on deep neural networks are responding to this need by developing tools for visualizing and interpreting the computations inside these vast structures. Such tools can visualize the activity on each layer of a trained network in real time, as it processes visual input (Yosinski, Clune, Nguyen, Fuchs, and Lipson, 2015). This

does help to gain intuition and learn something about the processes in the internal layers.

7. I am grateful for permission to base this section on my article "A Frame of Mind," originally published in the *Royal Society of Arts Journal*, 2, 2018, 46–47.

8. *The New York Times*, June 27, 2015, https://www.nytimes.com/2015/06/28/opinion/sunday/face-it-your-brain-is-a-computer.html. See also the book edited by Marcus and Jeremy Freeman, *The Future of the Brain*, Princeton University Press, 2014, https://press.princeton.edu/titles/10306.html.

9. Note that nondeterministic systems and probabilistic systems are included here, but the point is that any sufficiently large system can produce nondeterministic output.

10. This means *any* digital device—computers, processors, or even memory—can be constructed from a large supply of just one type of transistor. This is why we have cheap and powerful computers. It means that a complete computer processor can be fabricated on a silicon chip using just one kind of device repeated over and over. Mass production costs increase dramatically with the number of different types of transistors on a chip. A typical chip might have as many as 1 billion transistors, and if these are all identical, the manufacturing costs are hugely reduced.

11. In the latter case, very crude approximations may be entirely justified if, for example, statistical analyses show that lower tolerance in some of the details makes no significant difference in the results. To illustrate through an analogy: When investigating traffic flow in a city road system, the model of a typical vehicle can be very simple—perhaps just a few parameters for destination, speed, and direction might be sufficient—but many vehicle models will be used, perhaps thousands. In contrast, when studying road dynamics or collision behavior, much more elaborate models will be used, probably requiring full details of size, shape, materials, and other elements—yet far fewer of them will be needed (perhaps only one).

12. For example, GeNN is a popular neuronal network simulation environment based on Graphical Processing Units (GPUs) from NVIDIA (http://genn-team.github.io/genn/). See https://www.predictiveanalyticstoday.com/top-artificial-neural-network-software/ for a range of modeling tools, and https://grey.colorado.edu/emergent/index.php/Comparison_of_Neural_Network_Simulators for a comparison chart.

13. The "total connectivity" feature shows how AI designs can violate biological authenticity. It is very common in a multilayered design, like the one shown in figure 9.1, for each neuron to be connected to *every* neuron in the layer below. This has the advantage that all possible connections are available, in case they are required, and those that are not used will simply be given a weight of zero during learning. However, this design is totally unrealistic for the brain—there simply isn't room in the skull (Braitenberg, 2001). Not only is there not enough space for so many connections, but the proportion of unused links would be much higher than even the 50 percent pruning rates reported for early cortical development (Purves and Lichtman, 1980). Furthermore, this approach is also infeasible, as it would require an inordinately long learning period as all the weights are adjusted. But this objection

matters only if claims are made of a "brain model": for AI applications, it is of no concern (other than possibly increasing efficiency). Interestingly, one way around this particular issue might be the use of *small-world* network structures, which may exist in the brain. *Small worlds* are networks that have closely connected local groups with a few long-range links between the groupings. It was discovered in 1998 that the effect of total connectivity can be achieved in a small world network with a minimum number of steps. This is often described using analogies like "everyone is only six handshakes away from the president of the United States." See Humphries, Gurney, and Prescott (2006) for an example of small-world networks in brain models.

14. A brief summary of the many different types is provided at https://towards datascience.com/the-mostly-complete-chart-of-neural-networks-explained-3fb6f236 7464.

15. For example, the cascade-correlation method recruits new nodes (neurons) into the network layers to improve performance and removes less effective ones (Shultz, 2003, 2012). This "proliferation and pruning" process works by maintaining a pool of spare nodes and frequently checks to see if any of these could reduce the current error. If so, they are recruited into the main system (and weak nodes are correspondingly relegated). Another method of growing networks is the neural gas technique, which is so called because the network expands as new nodes are added and the growth looks like the molecular flow of a gas expanding within an enclosing space (Fritzke, 1995).

16. Personal communication with Christopher Maddison and David Silver from Google DeepMind, October 16, 2015.

17. Normally, the term *black box* means a device for which the internal workings are not visible. The black-box recorders used on airliners and other vehicles are crashproof data recorders (but they actually are painted bright orange or red, not black).

18. For example, changes made to an image that are so small that they are not noticeable to humans can cause a deep convolutional neural network (CNN) to misclassify the image as something completely different (Goodfellow, Shlens, and Szegedy, 2014). Also, images can be created from random dot patterns that are completely meaningless to humans but are recognized by a deep network (e.g., an animal) with very high confidence (Nguyen, Yosinski, and Clune, 2014). It seems that these errors are caused by the distributed nature of the patterns in such large networks—the images that we recognize as close to each other are not necessarily close inside the network (Szegedy et al., 2013).

19. For large neural networks, the combinatorial numbers of states of the many network parameters is many, many times more than the number of components in the network. Some form of modular structure would reduce this a bit.

20. See https://history.nasa.gov/sts1/pages/computer.html. Actually, it would be nice to see some conservatism from some of the software companies instead of them relying on users to find problems with the latest version release.

21. See "Verification," on Rich Sutton's home webpage: incompleteideas.net/.

22. *The Matrix* was written and directed by Lana and Andy Wachowski and released in 1999. In the film, most humans live in a simulated reality called "The Matrix," which is controlled by machines.

CHAPTER 10

1. These quotations are taken from Workshop 6, "Intelligent Robotic Systems," at the Twenty-Seventh AAAI Conference on Artificial Intelligence (given by the American Association for Artificial Intelligence) July 14–18, 2013, Bellevue, Washington. Six years later, the comments still stand.

2. See www.bostondynamics.com/robot_Atlas.html.

3. At the Tokyo Motor Show, on October 29, 2015, Yamaha revealed a robot that rode an unmodified high-speed racing motorcycle. (www.bbc.co.uk/news/technology-34665442). See also https://global.yamaha-motor.com/showroom/motobot/.

4. See https://global.toyota/en/detail/275807?_ga=2.137550000.1013375613.15642 54219-1463655293.1564254219.

5. This idea was developed in AI in the 1980s as a knowledge-based way of solving problems known as *blackboard systems*. A common area, the *blackboard*, would be updated by information from a group of specialist modules. As different modules shared their knowledge, pieces of the solution would eventually assemble on the blackboard.

6. See http://www.iclr.cc/.

7. Actually, general AI is the implicit (or sometimes explicit) long-term goal behind all AI research. Many researchers and scientists give passing mention of this long-term aim in the "Future Work" or "Discussion" sections of published papers, but usually nothing concrete is proposed.

8. See also Legg (2006) or Legg (2008) for full details.

9. D. Deutsch, *Creative blocks: The very laws of physics imply that artificial intelligence must be possible. What's holding us up?* Aeon, October 3, 2012; available from https://aeon.co/essays/how-close-are-we-to-creating-artificial-intelligence.

10. Karl Popper was a philosopher of science who argued that a theory based on scientific evidence can never be proven to be true but can be falsified by a single counterexample. This is an alternative to the empiricist approach, which equates the truth of a theory with the quantity of supporting evidence. It has resonances with software engineering, wherein software is never assumed to be correct and new bugs can always be found.

11. Deutsch (2012).

12. See http://agi-conf.org/2019/.

13. See https://Intelligence.org.

14. See https://deepmind.com/; note that content seems to change frequently on this site.

15. It's important to note that these are two-person, zero-sum games with no randomness or uncertainty. This is a small segment of the general class of games.

16. The OpenCog Foundation website is https://opencog.org/.

17. https://wiki.opencog.org/w/OpenCogPrime: FAQ on February 13 2017.

CHAPTER 11

1. In science fiction, anything seems possible, but in reality, there are many constraints on the universe we live in, and serious science-fiction authors like Arthur C. Clarke are careful to avoid the more obviously ridiculous scenarios.

2. This quote comes from a 1949 letter from the club founder and organizer, John Bates, a neurologist at the National Hospital for Nervous Diseases in London, where most of the club meetings were held. Bates organized and ran the Ratio Club, and Husband and Holland provide a good deal of interesting and important detail from the unpublished Bates archive (Husbands and Holland, 2008).

3. See the original proposal for the 1956 Dartmouth Summer Project on AI (McCarthy, Minsky, and Rochester, 1955).

4. This was before transistors. Electronics meant valves or tubes with heated elements that required lots of battery power.

5. Holland and Husbands have gathered together much information on all these early British cyberneticians (e.g., Husbands and Holland, 2012, and the excellent Husbands, Holland, and Wheeler, 2008). For an entertaining history of the origins and early growth of cybernetics and AI in the United States, see McCorduck (1979) and her slightly updated version of 2004.

6. Many of these machines are described in a book by the Hungarian engineer, inventor, and pioneer of color television, T. N. Nemes (1969).

7. Some examples include the Cybernetics Society (www.cybsoc.org/), the American Society for Cybernetics (asc-cybernetics.org/), and the World Organisation of Systems and Cybernetics (wosc.co/norbert-wiener-institute-of-systems-and-cybernetics/).

CHAPTER 12

1. Two previous conferences have combined into the key conference series: the Joint IEEE International Conference on Developmental Learning and Epigenetic Robotics (http://www.icdl-epirob.org/). The journal *IEEE Transactions on Cognitive and Developmental Systems (TCDS)* was previously entitled *IEEE Transactions on Autonomous Mental Development (TAMD)* (https://ieeexplore.ieee.org/xpl/RecentIssue.jsp?punumber=7274989).

2. See https://cis.ieee.org/technical-committees/cognitive-and-developmental-systems-technical-committee.

3. Constructivism also influences neural modeling; for descriptions and examples of the principles and practice of integrating development and brain modeling, see the

pair of volumes by Mareschal et al. (2007a) and Mareschal, Sirois, Westermann, and Johnson (2007b).

4. The nature versus nurture argument, or nativist versus empiricist stance refers to our lack of knowledge as to whether various brain functions are built in and determined by genetics or are learned through experience. It seems likely that a bit of both is involved in many cases, but the constructivist approach provides another way of looking at development and allows us to defer the debate until more understanding is gained.

5. Many infant studies do start at later developmental points, but they can carry a heavy obligation. The skills and competence of an infant at a particular time will depend profoundly upon her or his history of experience. For example, when building computer models of speech maturity that purport to show, say, how speech becomes related to images at the age of three years old, it is very important to understand and specify the skills and context gained from previous years. It might be very convenient to have subjects that can communicate with the experimenter, but if any of the knowledge and structures built up through prior years of extensive listening, verbal babbling, and proto-speech are ignored, then the assumptions at the starting point can be massively undermined. Of course, experimenters do strive hard to set up reasonable and accepted initial conditions, but work is still needed to explore the early stages so that the best and most reliable assumptions can be used.

6. Actually, there is an important distinction between sequences and stages. *Sequences* are seen when one skill develops before another, as in our timelines. *Stages* are ensembles of related developments that occur together and mark out a major developmental milestone. Although there is much debate about stages, sequences are noncontroversial. The distinction is not important for our purposes, and the terms are used interchangeably here.

7. As with all our timelines, this was constructed from many sources in the literature; see White, Castle, and Held (1964) for a good start on this table.

8. The importance and power of external constraint are sadly manifest in negative cases, such as when deprivation in childhood has lifelong effects.

9. This special issue of *Cognitive Science* contains five papers on various theories on constraints, as discussed in Keil (1990).

10. Be warned—language processing is a controversial area: for a refutation of Elman, see Douglas L. T. Rohde and David C. Plaut, "Less is less in language acquisition," in Quinlan (2003), 189–232.

11. Peter Gray gives five key characteristics of play in the article "Definitions of Play," at http://www.scholarpedia.org/article/Definitions_of_Play.

12. Piaget discusses key characteristics such as autotelism (deriving meaning from within), spontaneity (self-generated), and pleasure (when it stops being fun, they stop playing) (Piaget, 1999).

13. This relates to Hebbian learning, named after the psychologist Donald Hebb, in which related events become reinforced through repetition of their coexistence

(Caporale and Dan, 2008). The biochemical details of this are still not fully understood and are often updated.

CHAPTER 13

1. It is extremely difficult to live without proprioception. Some temporary local loss can occur during athletic activities, but extensive loss can be catastrophic—much worse than even blindness.

2. See either https://www.youtube.com/watch?v=OhWeKIyNcj8 or https://www.aber.ac.uk/en/cs/research/ir/robots/icub/dev-icub/.

3. Buttons could be programmed so that various lights would turn on and attract the iCub's attention according to the prevailing experimental agenda.

4. The whole of the cortical sheet, within the 3-mm thickness, consists of six distinct neural layers. Looking down vertically through the sheet, the neurons form groups, known as *cortical columns*, which are much more interconnected within the columns than between columns. This microstructure is repeated across the cortical sheet, and so it must be the basic functional unit of the cerebral cortex. This offers the tantalizing idea that if we could understand the microcircuit of just one such column, we would be a lot closer to knowing how the brain works. Many think the canonical cortical column performs some kind of prediction; Karl Friston suggests that it operates as a Bayesian predictor (Bastos et al., 2012).

5. We assume that the strength of a stimulus falls off in proportion to its distance from the center of a field. So in figure 13.8, the stimulus is very near one field center, and this gives a stronger response than the others. The ratio of the signal strengths can reconstruct the location in a target array, as shown.

6. This is because each field that overlaps the stimulus adds another measurement of its position, and thus increases the total accuracy. Thus, there is a trade-off between the number of fields used and the size of the elements. A large degree of overlap in the structure reduces both the number of elements required and the number of links between a pair of maps, thus learning can be very fast.

7. Other robotic work has also shown that starting with a low resolution in sensory and motor systems gives faster learning than starting with full precision (Gómez, Lungarella, Eggenberger Hotz, Matsushita, and Pfeifer, 2004).

8. Other robotics work has reported on staged development as an effective strategy for reducing DoFs (Lungarella and Berthouze, 2002; Sporns and Edelman, 1993), and on being beneficial for the emergence of stable patterns and helping to bootstrap later stages (Berthouze and Lungarella, 2004).

9. This would be without subscribing to any view on the origins of such knowledge, innate or learned.

10. For examples of investigations using dynamical fields and ANNs to pursue similar objectives to our own, see, respectively, Sandamirskaya and Storck (2014) and Muhammad and Spratling (2015), the latter using an autoencoder design, as mentioned in chapter 9.

CHAPTER 14

1. A team at the University of Parma in Italy did much pioneering work on mirror neurons and can claim to be the discoverers of the phenomena (Rizzolatti, Fadiga, Gallese, and Fogassi, 1996).

2. However, the manufacturer, ITT, is developing this ability in new versions (Hu, Eljaik, Stein, Nori, and Mombaur, 2016). See also www.youtube.com/watch?v=yIrepoNJyGc.

3. We carry around our ego space; for instance, consider reading a book or threading a needle.

4. This development is neatly captured in the title of a research paper: "Travel Broadens the Mind" (Campos et al., 2000).

5. The Shadow Dexterous Hand is an impressive example (see www.shadowrobot.com/products/dexterous-hand/).

6. Some people classify hunger as an intrinsic motivation. For our purposes, however, it can be left out for now. Such basic biological drivers can and do interact with cognitive functions, but we avoid this by assuming an undiminished energy supply.

7. The Meka robot is a mobile manipulator developed by the company Meka Robotics (which was acquired by Google X in 2013). It has two arms with dexterous hands, a mobile base, and an expressive head.

8. Respective examples of these include the use of cameras or ultrasonic devices to compensate for blindness, and the disturbing effects known as *phantom limbs*.

9. Kenneth Forbus is a pioneer of qualitative modeling of human cognitive processes. An introduction to his ideas can be found in Forbus (2018).

10. See Greenspan and Van Swinderen (2004) for experiments on the perceptual discrimination of the fruit fly.

11. Godfrey-Smith's book describes how the octopus evolved completely independently from humans, effectively becoming an intelligent alien.

12. Note the title of a recent paper: "Building Machines That Learn and Think Like People" (Lake, Ullman, Tenenbaum, and Gershman, 2017). This article contains commentary from many other researchers and is a useful resource.

CHAPTER 15

1. See http://www.popsci.com/darpa-robotics-challenge-was-bust-why-darpa-needs-try-again.

2. More information on this can be found on the Nvidia site (https://www.nvidia.com/en-gb/deep-learning-ai/solutions/).

3. See https://www.tensorflow.org/.

4. See https://berkeleyautomation.github.io/dex-net/.

5. See https://www.eu-robotics.net/robotics_league/news/press/consumer-service-robots-erl-competition-at-1st-lisbon.html.

6. DARPA's CALO project (whose name stands for "Cognitive Assistant that Learns and Organizes") to create an administration assistant, which spanned from 2003 to 2008, can be seen as the origins of Siri and the following boom in virtual assistants; see http://www.ai.sri.com/project/CALO.

CHAPTER 16

1. For product-centered predictions, see analyses such as Webb (2017).

2. The simulation model results formed the basis of the 1972 report *The Limits to Growth*, commissioned by the Club of Rome, an international think tank. This report, as well as books like *The Population Bomb* by Professor Paul R. Ehrlich of Stanford University (Ehrlich, 1968) (which warned of mass starvation in the 1970s and 1980s), generated a great deal of debate and alarm. See http://www.donellameadows.org/wp-content/userfiles/Limits-to-Growth-digital-scan-version.pdf.

3. See statistics from the International Federation of Robotics, given at www.ifr.org/worldrobotics/. Military robotics are excluded from these figures, as they relate to different economic demands.

4. For further data, see https://www.therobotreport.com.

5. See https://www.cbinsights.com/research/.

6. This information is available from https://www.cbinsights.com/research/report/ai-in-numbers-q2-2019/?utm_source=CB+Insights+Newsletter&utm_campaign=0aaf03fe92-newsletter_general_Thurs_20190725&utm_medium=email&utm_term=0_9dc0513989-0aaf03fe92-88859309.

7. See www.kindred.ai/.

8. See https://numenta.com/.

9. See https://openai.com/.

10. See https://www.goodai.com/.

11. See http://www.vicarious.com/.

12. *New Scientist* magazine, November 30, 2016, https://www.newscientist.com/article/2114748-google-translate-ai-invents-its-own-language-to-translate-with/.

13. Tesla have a solar panel plant named Gigafactory 2, and the company is working on developing more Gigafactories in Europe.

14. Details are given in Standard J3016_201609 at the Society of Automotive Engineers (SME; https://www.sae.org/).

15. This was from the IEEE Workshop on Autonomous Technologies and Their Societal Impact, held at the International Conference on Prognostics and Health Management, on June 20, 2016, in Ottawa, Canada.

16. The quotations here are taken from the 5th International Conference on Internet of Things: Systems, Management and Security, October 2018, in Valencia, Spain.

NOTES TO CHAPTER 17   335

17. For examples of analysis and design, see https://www.mentor.com/products/ele ctrical-design-software/capital/additional-automation/electrical-analysis and https:// www.autodesk.com/solutions/generative-design.

18. See http://standards.ieee.org/develop/indconn/ec/autonomous_systems.html.

19. See https://www.weforum.org/press/2019/05/world-economic-forum-inaugurat es-global-councils-to-restore-trust-in-technology.

20. The details are given in the white paper, "Industrial Strategy: Building a Britain Fit for the Future," November 27, 2017, (https://www.gov.uk/government/publica tions/industrial-strategy-building-a-britain-fit-for-the-future).

21. President's Lecture, at the Royal Society for the Encouragement of Arts, Manufactures, and Commerce, November 14, 2017, in London.

22. See https://deepmind.com/applied/deepmind-ethics-society/.

23. See https://opensource.org/.

24. The Select Committee on Artificial Intelligence was appointed by the House of Lords on June 29, 2017, "to consider the economic, ethical and social implications of advances in artificial intelligence." The report is available at www.parliament.uk /ai-committee.

25. This is available from the Westminster eForum Keynote Seminar, Artificial intelligence and Robotics: Innovation, Funding and Policy Priorities, held February 27, 2018, in London.

26. See *Artificial Intelligence, Automation, and the Economy*, Executive Office of the President, December 2016; available from https://obamawhitehouse.archives.gov /blog/2016/12/20/artificial-intelligence-automation-and-economy.

CHAPTER 17

1. I refer to our existing, working, hardware and software, not the theoretical science and complex problems behind their design and realization.

2. See Armstrong and Sotala (2015). For a recording of the talk, see https://www .youtube.com/watch?v=PwULqhOTm34.

3. A few cases did not properly address the questions asked and thus were ignored.

4. For example, Bostrom described a survey of 170 people asking, among other questions, when superintelligence is likely to be achieved (Bostrom, 2014, chapter 1). The participants from the largest group (42 percent of the total) expected superintelligence to occur much sooner than the others; they suggested a 90 percent probability of emergence within 30 years, while the other three groups clustered around 55 percent. This group included participants from a specialist conference on AGI, and so by actively working on the topic, they might be expected to have more belief in its future existence. They also had the strongest views on an extremely bad outcome for humanity. The full survey details are in a later paper (Müller and Bostrom, 2016). (The questionnaire asked about HLMI, *high-level machine intelligence*, while the book

uses the term *human-level machine intelligence*, these can be taken to mean the same thing.)

5. If you think I'm skeptical about this topic, I'm joined by plenty of others. David Deutsch says: 'I cannot think of any other significant field of knowledge [AGI] in which the prevailing wisdom, not only in society at large but also among experts, is so beset with entrenched, overlapping, fundamental errors. Yet it has also been one of the most self-confident fields in prophesying that it will soon achieve the ultimate breakthrough.' From *Creative Blocks: The Very Laws of Physics Imply That Artificial Intelligence Must Be Possible. What's Holding Us Up?* Aeon, October 3, 2012; available from https://aeon.co/essays/how-close-are-we-to-creating-artificial-intelligence.

6. Computer programs can generate random programs and then run and evolve them until they achieve some goal. In theory, it is possible to produce *any* program this way, just like the monkeys eventually typing a Shakespearian sonnet. However, even if such an experiment did run for a long time and produced some kind of malignant superintelligence, the machine would still be under the control of the programmer, it would still be in a laboratory, and the experimenter still has the power to turn off the machine. The most likely outcome would be that the program would be analyzed under controlled conditions until its function was understood.

7. Bostrom argues that it does not need emotional drives such as anger or revenge; the goals could arise as an optimization of resources (e.g., electricity is needed for machines, not humans). He also suggests that superintelligence systems would build up their plans in secret before attacking and diverting resources. But this ignores the thousands of software experts, information technology (IT) managers, amateurs, users, and hackers who directly or indirectly monitor the patterns of activity of the world's computers. The idea that a global network of computers might hide part of their processing or pretend that they are working on something else, and not be detected by *anyone*, is absurd. As soon as even one was detected, they would all be investigated, and remedial action would be started.

8. The IoT will make it easy to control all kinds of devices and services remotely (and will bring enormous scope for abuse), but to achieve total, dictatorial control, you need high-quality, reliable command paths to all utilities, resources, and assets. And these always involve humans in the chain somewhere.

9. Unfortunately, even some politicians engage in science-bashing, saying that academics and other experts have no experience of the "real world." They don't see that expertise and experience are on the same spectrum—long and deep experience of a topic is the source of expertise. And as for the real world, do they think that financial managers, politicians, and vociferous bloggers have a better, more correct grasp of the world?

10. Karl Popper produced a classic book on this (Popper, Ryan, and Gombrich, 2013).

11. For example, if a skull image was flashed up during a horror film, then people would find the film more frightening than would a control group who didn't see the image. Similarly, if some delicious food was the subject of a flashed image, the audience would report feeling hungry.

12. See https://www.ted.com/talks/hans_rosling_shows_the_best_stats_you_ve_ever_seen?language=en.

13. Of course, there are still problems. While the proportion of people living in extreme poverty is nearly half of what it was two decades ago, the wealth of the very rich is increasing. Apparently, the richest 10 percent of the world now own 88 percent of its wealth.

14. In the annual report, *State of the Climate in 2017*, by the American Meteorological Society, 2015, 2016, and 2017 are cited as the top three warmest years for global temperature since records began in 1850. In fact, 16 of the warmest 17 years have occurred since the year 2000. All the scientific data gives a strong consensus that global warming is not only a possible but a highly probable event, with very serious consequences.

15. If both Greenland and Antarctica lost all their land ice, sea levels would rise by 200 ft.

16. See https://www.gsb.stanford.edu/insights/andrew-ng-why-ai-new-electricity.

17. See https://www.turing.ac.uk/.

18. This appears "Rise of the Machines: Has Technology Evolved beyond Our Control?" *The Guardian*, June 15, 2018; available at https://www.theguardian.com/books/2018/jun/15/rise-of-the-machines-has-technology-evolved-beyond-our-control. This idea is further developed in Bridle's book, *New Dark Age* (Bridle, 2018).

BIBLIOGRAPHY

Ariely, Dan. *Predictably irrational*. HarperCollins, 2008.

Armstrong, Stuart, and Sotala, Kaj. "How we're predicting AI—or failing to." In J. Romportl, E. Zackova, and J. Kelemen, eds., *Beyond artificial intelligence: The disappearing human-machine divide* (11–29). Springer, 2015.

Ashby, W. Ross. *Design for a brain: The origin of adaptive behaviour*. 2nd ed. Chapman and Hall, 1976.

Banavar, Guruduth. *Learning to trust artificial intelligence systems: Accountability, compliance and ethics in the age of smart machines, 2016*. IBM technical report. Available from https://www.alain-bensoussan.com/wp-content/uploads/2017/06/34348524.pdf.

Barlow, Horace B. "Cerebral cortex as model builder." In D. Rose and V. G. Dobson eds., *Models of the visual cortex* (37–46). John Wiley, 1985.

Bastos, Andre M., Usrey, W. Martin, Adams, Rick A., Mangun, George R., Fries, Pascal, and Friston, Karl J. "Canonical microcircuits for predictive coding." *Neuron*, 76(4), 695–711, 2012.

Becker, J. D. "A model for the encoding of experiental information." In R. C. Schank and K. M. Colby, eds., *Computer models of thought and language* (396–434). W. H. Freeman and Company, 1973.

Bengio, Yoshua, Courville, Aaron, and Vincent, Pierre. "Representation learning: A review and new perspectives." *IEEE Transactions on Pattern Analysis and Machine Intelligence*, 35(8), 1798–1828, 2013.

Bernstein, Nikolai A. *The co-ordination and regulation of movements*. Pergamon Press Ltd., 1967.

Berthouze, L., and Lungarella, M. "Motor skill acquisition under environmental perturbations: On the necessity of alternate freezing and freeing of degrees of freedom." *Adaptive Behavior*, 12(1), 47–64, 2004.

Bessen, James E. *How computer automation affects occupations: Technology, jobs, and skills*. Technical Report No. 15-49, Boston University, School of Law, Law and Economics Research Paper, 2016.

Bizony, Piers. *The making of Stanley Kubrick's 2001: A Space Odyssey*. Taschen, 2015.

Blum, Christian, Winfield, Alan F. T., and Hafner, Verena V. "Simulation-based internal models for safer robots." *Frontiers in Robotics and AI*, 4, 74, 2018. Retrieved from https://www.frontiersin.org/articles/10.3389/frobt.2017.00074/full.

Boden, Margaret, Bryson, Joanna, Caldwell, Darwin, et al. "Principles of robotics: regulating robots in the real world." *Connection Science*, 29(2), 124–129, 2017.

Boden, Margaret A. *AI: Its nature and future*. Oxford University Press, 2016.

Borghi, Anna M., Flumini, Andrea, Natraj, Nikhilesh, and Wheaton, Lewis A. "One hand, two objects: emergence of affordance in contexts." *Brain and Cognition*, 80(1), 64–73, 2012.

Bostrom, Nick. *Superintelligence: Paths, dangers, strategies*. Oxford University Press, 2014.

Bottou, Léon. "From machine learning to machine reasoning." *Machine Learning*, 94(2), 133–149, 2014.

Braitenberg, Valentino. "Brain size and number of neurons: An exercise in synthetic neuroanatomy." *Journal of Computational Neuroscience*, 10(1), 71–77, 2001.

Braud, Raphael, Giagkos, Alexandros, Shaw, Patricia, Lee, Mark, and Shen, Qiang. "Building representations of proto-objects with exploration of the effect on fixation times." In *7th International Conference on Development and Learning and on Epigenetic Robotics, ICDL-EpiRob 2017*, 296–303. 2017.

Breazeal, Cynthia L. *Designing sociable robots*. MIT Press, 2002.

Bridle, James. *New dark age: Technology and the end of the future*. Verso Books, 2018.

Brockman, John. *What to think about machines that think*. Harper Perennial, 2015.

Brooks, Rodney A. "Intelligence without representation." *Artificial Intelligence*, 47(1–3), 139–159, 1991.

Bruinsma, Anne Hendrik. *Practical robot circuits: Electronic sensory organs and nerve systems*. Philips Technical Library, 1960.

Bruner, J. S., Jolly, A., and Sylva, K. *Play: Its role in development and evolution*. Basic Books, 1976.

Bruner, Jerome Seymour. *Toward a theory of instruction*. Harvard University Press, 1966.

Bruner, Jerome Seymour. *Acts of meaning*. Harvard University Press, 1990.

Bushnell, Emily W., and Boudreau, J. Paul. "Motor development and the mind: The potential role of motor abilities as a determinant of aspects of perceptual development." *Child Development*, 64(4), 1005–1021, 1993.

Busso, Carlos, Deng, Zhigang, Yildirim, Serdar, et al. "Analysis of emotion recognition using facial expressions, speech and multimodal information." In *Proceedings of the 6th International Conference on Multimodal Interfaces*, 205–211. ACM, 2004.

Campbell, Robert L., and Bickhard, Mark H. "Types of constraints on development: An interactivist approach." *Developmental Review*, 12(3), 311–338, 1992.

Campos, Joseph J., Anderson, David I., Barbu-Roth, Marianne A., Hubbard, Edward M., Hertenstein, Matthew J., and Witherington, David. "Travel broadens the mind." *Infancy*, 1(2), 149–219, 2000.

Cangelosi, A., Metta, G., Sagerer, G., et al. "Integration of action and language knowledge: A roadmap for developmental robotics." *IEEE Transactions on Autonomous Mental Development*, 2(3), 167–195, 2010.

Cangelosi, Angelo, and Schlesinger, Matthew. *Developmental robotics: From babies to robots*. MIT Press, 2015.

Caporale, Natalia, and Dan, Yang. "Spike timing-dependent plasticity: a Hebbian learning rule." *Annual Review of Neuroscience*, 31, 25–46, 2008.

Charness, Neil. "Search in chess: Age and skill differences." *Journal of Experimental Psychology: Human Perception and Performance*, 7(2), 467, 1981.

Chatila, Raja, Renaudo, Erwan, Andries, Mihai, et al. "Toward self-aware robots." *Frontiers in Robotics and AI*, 5, 88, 2017.

Clark, Andy. *Being there—Putting brain, body, and world together again*. MIT Press, 1998.

Clark, Andy. "Whatever next? Predictive brains, situated agents, and the future of cognitive science." *Behavioral and Brain Sciences*, 36(3), 181–204, 2013.

Cohen, Jack, and Stewart, Ian. *The collapse of chaos: Discovering simplicity in a complex world*. Penguin UK, 1995.

Craik, Kenneth J. W. *The nature of explanation*. Cambridge University Press, 1943. Reprinted 1967.

Crick, Francis. *The astonishing hypothesis*. Simon and Schuster, 1994.

Damasio, Antonio. *The feeling of what happens: Body and emotion in the making of consciousness*. Vintage, 2000.

Daugherty, Paul R., and Wilson, H. James. *Human + machine: Reimagining work in the age of AI*. Harvard Business Review Press, 2018.

Dautenhahn, Kerstin. "Socially intelligent robots: dimensions of human–robot interaction." *Philosophical Transactions of the Royal Society of London B: Biological Sciences*, 362(1480), 679–704, 2007.

Davis, Ernest, and Marcus, Gary. "The scope and limits of simulation in automated reasoning." *Artificial Intelligence*, 233, 60–72, 2016.

Davis, Josh P., Lander, Karen, Evans, Ray, and Jansari, Ashok. "Investigating predictors of superior face recognition ability in police super-recognisers." *Applied Cognitive Psychology*, 30(6), 827–840, 2016.

de Garis, Hugo, Shuo, Chen, Goertzel, Ben, and Ruiting, Lian. "A world survey of artificial brain projects, part 1: Large-scale brain simulations." *Neurocomputing*, 74(1–3), 3–29, 2010.

de Groot, Adriaan D. *Thought and choice in chess*, vol. 4. de Gruyter, 2014.

de Latil, Pierre. *Thinking by machine: A study of cybernetics*. Trans. Y. M. Golla. Sidgwick and Jackson, 1956.

Dennett, Daniel C. "Reflections on language and mind." In Peter Carruthers and Jill Boucher, eds., *Language and thought: Interdisciplinary themes*, 284–294, Cambridge University Press, 1998.

Denning, Peter J., ed. *Talking back to the machine: Computers and human aspiration*. Springer-Verlag, 1999.

Der, Ralf, and Martius, Georg. *The playful machine—Theoretical foundation and practical realization of self-organizing robots*. Cognitive Systems Monographs (COSMOS). Springer-Verlag, 2011.

Derdikman, D., and Moser, E. I. "A manifold of spatial maps in the brain." *Trends in Cognitive Sciences*, 14(12), 561–569, 2010.

de Vries, Johanna I. P., Visser, Gerard H. A., and Prechtl, Heinz F. R. "The emergence of fetal behaviour. I. Qualitative aspects." *Early Human Development*, 7(4), 301–322, 1982.

de Waal, Frans. *Are we smart enough to know how smart animals are?* W. W. Norton & Company, 2016.

Dickmanns, Ernst Dieter. *Dynamic vision for perception and control of motion*. Springer Science & Business Media, 2007.

Domingos, Pedro. *The master algorithm: How the quest for the ultimate learning machine will remake our world*. Basic Books, 2015.

Dorigo, Marco, and Colombetti, Marco. *Robot shaping: An experiment in behavior engineering*. MIT Press, 1998.

Drescher, Gary L. *Made up minds: A constructivist approach to artificial intelligence.* MIT Press, 1991.

Earland, Kevin, Lee, Mark, Shaw, Patricia, and Law, James. "Overlapping structures in sensory-motor mappings." *PloS ONE,* 9(1), e84240, 2014.

Edelman, Gerald M. *Bright air, brilliant fire: On the matter of the mind.* Basic Books, 1992.

Ehrlich, Paul R. *The population bomb.* Sierra Club/Ballantine Books, 1968.

Elman, Jeffrey L. "Learning and development in neural networks: The importance of starting small." *Cognition,* 48(1), 71–99, 1993.

Fauconnier, Gilles. *Mappings in thought and language.* Cambridge University Press, 1997.

Flavell, J. H. *The developmental psychology of Jean Piaget.* van Nostrand, 1963.

Fong, Terrence, Nourbakhsh, Illah, and Dautenhahn, Kerstin. "A survey of socially interactive robots." *Robotics and Autonomous Systems,* 42(3–4), 143–166, 2003.

Forbus, Kenneth D. *Qualitative representations: How people reason and learn about the continuous world.* MIT Press, 2018.

Forestier, Sébastien, and Oudeyer, Pierre-Yves. "Modular active curiosity-driven discovery of tool use." In *2016 IEEE/RSJ International Conference on Intelligent Robots and Systems (IROS),* 3965–3972. IEEE, 2016.

Frase, Peter. *Four futures: Life after capitalism.* Verso Books, 2016.

Fritzke, Bernd. "A growing neural gas network learns topologies." *Advances in Neural Information-Processing Systems,* 7, 625–632, 1995.

Gallagher, Shaun. "Philosophical conceptions of the self: implications for cognitive science." *Trends in Cognitive Sciences,* 4(1), 14–21, 2000.

Gallese, Vittorio. "Embodied simulation: From neurons to phenomenal experience." *Phenomenology and the Cognitive Sciences,* 4(1), 23–48, 2005.

Gates, Bill. "A robot in every home." *Scientific American,* 296(1), 58–65, 2007.

Gerson, Sarah A., and Woodward, Amanda L. "The joint role of trained, untrained, and observed actions at the origins of goal recognition." *Infant Behavior and Development,* 37(1), 94–104, 2014.

Giagkos, A., Lewkowicz, D., Shaw, P., Kumar, S., Lee, M., and Shen, Q. "Perception of localized features during robotic sensorimotor development." *IEEE Transactions on Cognitive and Developmental Systems,* 9(2), 127–140, 2017; doi:10.1109/TCDS.2017.2652129.

Ginsburg, K. R., and the Committee on Communications, and the Committee on Psychosocial Aspects of Child and Family Health. "The importance of play in promoting healthy child development and maintaining strong parent-child bonds." *Pediatrics*, 119(1), 182, 2007.

Gobet, Fernand. *Understanding expertise: A multi-disciplinary approach*. Macmillan International Higher Education, 2015.

Godfrey-Smith, Peter. *Other minds: The octopus and the evolution of intelligent life*. William Collins, 2017.

Goldin-Meadow, Susan. "The role of gesture in communication and thinking." *Trends in Cognitive Sciences*, 3(11), 419–429, 1999.

Gómez, Gabriel, Lungarella, Max, Eggenberger Hotz, Peter, Matsushita, Kojiro, and Pfeifer, Rolf. "Simulating development in a real robot: on the concurrent increase of sensory, motor, and neural complexity." In L. Berthouze, H. Kozima, C. G. Prince, et al. eds., *Proceedings of the Fourth International Workshop on Epigenetic Robotics*, Lund University Cognitive Studies, 117, 1–24 2004.

Good, Irving John. "Speculations concerning the first ultraintelligent machine." *Advances in Computers*, 6(31), 31–88, 1965.

Goodfellow, Ian J., Shlens, Jonathon, and Szegedy, Christian. "Explaining and harnessing adversarial examples." *arXiv preprint, arXiv:1412.6572*, 2014.

Gottlieb, Jacqueline, Oudeyer, Pierre-Yves, Lopes, Manuel, and Baranes, Adrien. "Information-seeking, curiosity, and attention: Computational and neural mechanisms." *Trends in Cognitive Sciences*, 17(11), 585–593, 2013.

Grand, Steve. *Growing up with Lucy: How to Build an Android in Twenty Easy Steps*. Phoenix, 2004.

Greenspan, Ralph J., and Van Swinderen, Bruno. "Cognitive consonance: complex brain functions in the fruit fly and its relatives." *Trends in Neurosciences*, 27(12), 707–711, 2004.

Hamann, Katharina, Warneken, Felix, Greenberg, Julia R., and Tomasello, Michael. "Collaboration encourages equal sharing in children but not in chimpanzees." *Nature*, 476(7360), 328–331, 2011.

Haque, Usman. "The architectural relevance of Gordon Pask." *Architectural Design*, 77(4), 54–61, 2007.

Harari, Yuval Noah. *Homo Deus: A brief history of tomorrow*. Random House, 2016.

Harnad, Stevan. "The symbol grounding problem." *Physica D: Nonlinear Phenomena*, 42(1), 335–346, 1990.

Hawkins, Jeff, and Blakeslee, Sandra. *On intelligence: How a new understanding of the brain will lead to the creation of truly intelligent machines*. Macmillan, 2007.

Hoffmann, Matej, Marques, Hugo, Arieta, Alejandro, Sumioka, Hidenobu, Lungarella, Max, and Pfeifer, Rolf. "Body schema in robotics: a review." *IEEE Transactions on Autonomous Mental Development*, 2(4), 304–324, 2010.

Holland, Owen. "Exploration and high adventure: the legacy of Grey Walter." *Philosophical Transactions of the Royal Society of London. Series A: Mathematical, Physical, and Engineering Sciences*, 361(1811), 2085–2121, 2003.

Hsu, Feng-Hsiung. *Behind Deep Blue: Building the computer that defeated the world chess champion*. Princeton University Press, 2004.

Hu, Huosheng. "Biologically inspired design of autonomous robotic fish at Essex." In *Proceedings of the IEEE SMC UK-RI Chapter Conference on Advances in Cybernetic Systems,,* 1–8. 2006.

Hu, Yue, Eljaik, Jorhabib, Stein, Kevin, Nori, Francesco, and Mombaur, Katja D. "Walking of the iCub humanoid robot in different scenarios: Implementation and performance analysis." *Computing Research Repository, abs/1607.08525*, 2016.

Hubel, David H., and Wiesel, Torsten N. "Receptive fields, binocular interaction and functional architecture in the cat's visual cortex." *Journal of Physiology*, 160(1), 106, 1962.

Hughes, M., and Hutt, C. "Heart-rate correlates of childhood activities: play, exploration, problem-solving and day-dreaming." *Biological Psychology*, 8(4), 253–263, 1979.

Humby, Clive, Hunt, Terry, and Phillips, Tim. *Scoring points: How Tesco continues to win customer loyalty*. Kogan Page Publishers, 2008.

Humphries, Mark D., Gurney, Kevin, and Prescott, Tony J. "The brainstem reticular formation is a small-world, not scale-free, network." *Proceedings of the Royal Society of London B: Biological Sciences*, 273(1585), 503–511, 2006.

Hunnius, Sabine, and Bekkering, Harold. "What are you doing? How active and observational experience shape infants' action understanding." *Philosophical Transactions of the Royal Society B: Biological Sciences*, 369(1644), 20130490, 2014.

Husbands, Phil, and Holland, Owen. "Warren McCulloch and the British cyberneticians." *Interdisciplinary Science Reviews*, 37(3), 237–253, 2012.

Husbands, Philip, and Holland, Owen. "The Ratio Club: A hub of British cybernetics." In Philip Husbands, Owen Holland, and Michael Wheeler, eds., *The mechanical mind in history*, 91–148. MIT Press, 2008.

Husbands, Philip, Holland, Owen, and Wheeler, Michael, eds. *The mechanical mind in history*. MIT Press, 2008.

Hutchins, Edwin. *Cognition in the wild*. MIT Press, 1995.

Jamone, Lorenzo, Bernardino, Alexandre, and Santos-Victor, José. "Benchmarking the grasping capabilities of the iCub hand with the YCB object and model set." *IEEE Robotics and Automation Letters*, 1(1), 288–294, 2016.

Johnson, M., Schuster, M., Le, Q. V., et al. "Google's multilingual neural machine translation system: Enabling zero-shot translation." *arXiv e-prints, arXiv:1611.04558*, 2016.

Johnson, Mark H., and de Haan, Michelle. *Developmental cognitive neuroscience: An introduction*. 4th ed. John Wiley & Sons, 2015.

Kaas, Jon H. "Topographic maps are fundamental to sensory processing." *Brain Research Bulletin*, 44(2), 107–112, 1997.

Kahneman, Daniel. *Thinking, fast and slow*. Penguin Books, 2012.

Karmiloff-Smith, Annette. *Beyond modularity: A developmental perspective on cognitive science*. MIT Press, 1995.

Kasparov, Garry. *Deep thinking: Where machine intelligence ends and human creativity begins*. John Murray, 2017.

Keil, Frank C. "Constraints on constraints: Surveying the epigenetic landscape." *Cognitive Science*, 14(1), 135–168, 1990.

Keysers, Christian, and Gazzola, Valeria. "Social neuroscience: Mirror neurons recorded in humans." *Current Biology*, 20(8), R353–R354, 2010.

King, Elizabeth. "Clockwork prayer: A sixteenth-century mechanical monk." *Blackbird: An Online Journal of Literature and the Arts*, 1(1), 1–29, 2002.

Kirby, Rachel, Forlizzi, Jodi, and Simmons, Reid. "Affective social robots." *Robotics and Autonomous Systems*, 58(3), 322–332, 2010.

Klatzky, Roberta L. "Allocentric and egocentric spatial representations: Definitions, distinctions, and interconnections." In C. Freksa, C. Habel, and K. F. Wender, eds., *Spatial cognition. Lecture Notes in Computer Science* 1404, 1–17. Springer, 1998.

Kober, Jens, Bagnell, J. Andrew, and Peters, Jan. "Reinforcement learning in robotics: A survey." *International Journal of Robotics Research*, 32(11), 1238–1274, 2013.

Kozima, Hideki, Michalowski, Marek P., and Nakagawa, Cocoro. "Keepon: A playful robot for research, therapy, and entertainment." *International Journal of Social Robotics*, 1(1), 3–18, 2009.

Krizhevsky, Alex, Sutskever, Ilya, and Hinton, Geoffrey E. "ImageNet classification with deep convolutional neural networks." In P. Bartlett, F. C. N. Pereira, C. J. C. Burges, L. Bottou, and K. Q. Weinberger, eds., *Advances in neural information processing systems*, 1097–1105. Curran Associates, 2012.

Kumar, Suresh, Shaw, Patricia, Giagkos, Alexandros, Braud, Raphael, Lee, Mark, and Shen, Qiang. "Developing hierarchical schemas and building schema chains through practice play behavior." *Frontiers in Neurorobotics*, 12(33), 2018; doi:10.3389/fnbot.2018.00033.

Kuniyoshi, Y., Yorozu, Y., Suzuki, S., et al. "Emergence and development of embodied cognition: A constructivist approach using robots." *Progress in Brain Research*, 164, 425–445, 2007.

Kuniyoshi, Yasuo, and Sangawa, Shinji. "Early motor development from partially ordered neural-body dynamics: experiments with a cortico-spinal-musculo-skeletal model." *Biological Cybernetics*, 95(6), 589–605, 2006.

Lake, Brenden M., Lawrence, Neil D., and Tenenbaum, Joshua B. "The emergence of organizing structure in conceptual representation." Special Issue: Memory, Learning, and Expertise. *Cognitive Science*, 42(S3), 809–832, 2018.

Lake, Brenden M., Ullman, Tomer D., Tenenbaum, Joshua B., and Gershman, Samuel J. "Building machines that learn and think like people." *Behavioral and Brain Sciences*, 40, e253, 2017; doi:10.1017/S0140525X16001837.

Lakoff, George. *Women, fire, and dangerous things: What categories reveal about the mind.* University of Chicago Press, 1987.

Law, James, Lee, Mark, Hülse, Martin, and Tomassetti, Alessandra. "The infant development timeline and its application to robot shaping." *Adaptive Behavior*, 19(5), 335–358, 2011.

Law, James A., Shaw, Patricia H., Earland, Kevin G., Sheldon, Michael T., and Lee, Mark H. "A psychology based approach for longitudinal development in cognitive robotics." *Frontiers in Neurorobotics*, 8(1), 1–19, 2014a; doi:10.3389/fnbot.2014.00001.

Law, James A., Shaw, Patricia H., and Lee, Mark H. "A biologically constrained architecture for developmental learning of eye-head gaze control on a humanoid robot." *Autonomous Robots*, 35(1), 77–92, 2013. doi:10.1007/s10514-013-9335-2.

Law, James A., Shaw, Patricia H., Lee, Mark H., and Sheldon, Michael T. "From saccades to grasping: A model of coordinated reaching through simulated development on a humanoid robot." *IEEE Transactions on Autonomous Mental Development*, 6(2), 93–109, 2014b; doi:10.1109/TAMD.2014.2301934.

LeCun, Yann, Bengio, Yoshua, and Hinton, Geoffrey. "Deep learning." *Nature*, volume 521(7553), 436–444, 2015.

LeCun, Yann, Bottou, Léon, Bengio, Yoshua, and Haffner, Patrick. "Gradient-based learning applied to document recognition." *Proceedings of the IEEE*, 86(11), 2278–2324, 1998.

Lee, Mark. "Intrinsic activity: from motor babbling to play." In *Prooceedings of the IEEE Joint International Conference on Development and Learning (ICDL) and Epigenetic Robotics (EpiRob)* 2, 1–6, 2011.

Lee, Mark, Law, James, and Hülse, Martin. "A developmental framework for cumulative learning robots." In *Computational and robotic models of the hierarchical organization of behavior*, 177–212. Springer, 2013.

Legg, Shane. "Is there an elegant universal theory of prediction?" In *Algorithmic Learning Theory, 17th International Conference*, 274–287. Springer, 2006.

Legg, Shane. *Machine super intelligence*. Ph.D thesis, University of Lugano, Switzerland, 2008.

Legg, Shane, and Hutter, Marcus. "A collection of definitions of intelligence." In B. Goertzel and P. Wang, eds., *Advances in artificial general intelligence*, vol. 157, 17–24. IOS Press, 2007.

Lepora, Nathan F., Martinez-Hernandez, Uriel, Evans, Mathew, Natale, Lorenzo, Metta, Giorgio, and Prescott, Tony J. "Tactile superresolution and biomimetic hyperacuity." *IEEE Transactions on Robotics*, 31(3), 605–618, 2015.

Levesque, Hector J., Davis, Ernest, and Morgenstern, Leora. "The Winograd Schema Challenge." In *13th International Conference on the Principles of Knowledge Representation and Reasoning, KR 2012*, 552–561. 2012.

Levine, Sergey, Pastor, Peter, Krizhevsky, Alex, and Quillen, Deirdre. "Learning hand-eye coordination for robotic grasping with Deep Learning and large-scale data collection." *Computing Research Repository (CoRR), arXiv:1603.02199*, 2016.

Lewis, Marc D. "The promise of dynamic systems approaches for an integrated account of human development." *Child Development*, 71(1), 36–43, 2000.

Lewis, Michael. *The Undoing Project: A friendship that changed our minds*. W. W. Norton, 2016.

Long, John. *Darwin's devices: What evolving robots can teach us about the history of life and the future of technology*. Basic Books, 2012.

Lungarella, M., and Berthouze, L. "Adaptivity through physical immaturity." In *Proceedings of the Second International Workshop on Epigenetic Robotics*, Lund University Cognitive Studies, 94, 79–86, 2002.

Lungarella, Max, Metta, Giorgio, Pfeifer, Rolf, and Sandini, Giulio. "Developmental robotics: a survey." *Connection Science*, 15(4), 151–190, 2003.

MacKay, Donald M. "Mind-like behaviour in artefacts." *Bulletin of the British Society for the History of Science*, 1(S5), 164–165, 1951.

MacKay, Donald M. "On comparing the brain with machines." *American Scientist*, 261–268, 1954.

MacKay, Donald M. *Information, mechanism, and meaning*. MIT Press, 1969.

Maddison, Chris J., Huang, Aja, Sutskever, Ilya, and Silver, David. "Move evaluation in go using deep convolutional neural networks." *arXiv preprint arXiv:1412.6564*, 2014.

Mahler, Jeffrey, Liang, Jacky, Niyaz, Sherdil, et al. "Dex-net 2.0: Deep learning to plan robust grasps with synthetic point clouds and analytic grasp metrics." *arXiv preprint arXiv:1703.09312*, 2017.

Mareschal, Denis, Johnson, Mark H., Sirois, Sylvain, Spratling, Michael, Thomas, Michael S. C., and Westermann, Gert. *Neuroconstructivism vol. I: How the brain constructs cognition*. Oxford University Press, 2007a.

Mareschal, Denis, Sirois, Sylvain, Westermann, Gert, and Johnson, Mark H. *Neuroconstructivism vol. II: Perspectives and prospects*. Oxford University Press, 2007b.

Marino, Lori. "Thinking chickens: A review of cognition, emotion, and behavior in the domestic chicken." *Animal Cognition*, 20, 127–147, 2017; doi:10.1007/s10071-016-1064-4.

Mason, Paul. *Postcapitalism: A guide to our future*. Macmillan, 2016.

Matin, Leonard, and Li, Wenxun. "Multimodal basis for egocentric spatial localization and orientation." *Journal of Vestibular Research: Equilibrium & Orientation*, 5(6), 499–518, 1995.

Maturana, Humberto R., and Varela, Francisco J. *Autopoiesis and cognition: The realization of the living*. Springer Science & Business Media, 1991.

McCarthy, J., Minsky, M. L., and Rochester, N. *A proposal for the Dartmouth Summer Research Project on Artificial Intelligence*. Technical report, Dartmouth College, Hanover, NH, available from http://jmc.stanford.edu/articles/dartmouth/dartmouth.pdf, 1955.

McColl, Derek, Hong, Alexander, Hatakeyama, Naoaki, Nejat, Goldie, and Benhabib, Beno. "A survey of autonomous human affect detection methods for social robots engaged in natural HRI." *Journal of Intelligent & Robotic Systems*, 82(1), 101–133, 2016.

McCorduck, Pamela. *Machines who think*. W. H. Freeman & Co., 1979, and 2nd edition 2004.

McCulloch, Warren S., and Pitts, Walter H. "A logical calculus of the ideas immanent in nervous activity." *Bulletin of Mathematical Biophysics* 5, 115–133, 1943.

Metta, G., Sandini, G., Vernon, D., Natale, L., and Nori, F. "The iCub humanoid robot: An open platform for research in embodied cognition." In *Proceedings of the 8th Workshop on Performance Metrics for Intelligent Systems*, ACM, 50–56. ACM, 2008.

Miikkulainen, Risto, Bednar, James A., Choe, Yoonsuck, and Sirosh, Joseph. *Computational maps in the visual cortex*. Springer Nature, 2006.

Minsky, Marvin L. "Logical versus analogical or symbolic versus connectionist or neat versus scruffy." *AI Magazine*, 12(2), 34–51, 1991.

Mnih, Volodymyr, Kavukcuoglu, Koray, Silver, David, et al. "Human-level control through deep reinforcement learning." *Nature*, 518(7540), 529–533, 2015.

Moorehead, Alan. *The Fatal impact: An account of the invasion of the South Pacific, 1767–1840*. Harper & Row, 1966.

Mori, Masahiro. "The uncanny valley." Translated by Karl F. MacDorman and Takashi Minato. *Energy*, 7(4), 33–35, 1970.

Morton, A. Jennifer, and Avanzo, Laura. "Executive decision-making in the domestic sheep." *PLoS ONE*, 6(1), e15752, 2011.

Muhammad, Wasif, and Spratling, Michael W. "A neural model of binocular saccade planning and vergence control." *Adaptive Behavior*, 23(5), 265–282, 2015.

Muhammad, Wasif, and Spratling, Michael W. "A neural model of coordinated head and eye movement control." *Journal of Intelligent & Robotic Systems*, 85(1), 107–126, 2017.

Müller, Vincent C., and Bostrom, Nick. "Future progress in artificial intelligence: A survey of expert opinion." In Vincent C Müller, ed., *Fundamental issues of artificial intelligence*, 376, 555–572. Springer, 2016.

Natale, Lorenzo, Nori, Francesco, Metta, Giorgio, et al. "The iCub platform: A tool for studying intrinsically motivated learning." In Gianluca Baldassarre and Marco Mirolli, eds., *Intrinsically motivated learning in natural and artificial systems*, 433–458. Springer Berlin Heidelberg, 2013; doi:10.1007/978-3-642-32375-1_17.

Nemes, T. N. *Cybernetic machines*. Iliffe Books Ltd., 1969. Translated from the 1962 Hungarian edition by I. Foldes.

Nguyen, Anh, Yosinski, Jason, and Clune, Jeff. "Deep neural networks are easily fooled: High confidence predictions for unrecognizable images." *arXiv preprint arXiv:1412.1897*, 2014.

O'Neil, Cathy. *Weapons of math destruction: How Big Data increases inequality and threatens democracy*. Crown Publishing Group (NY), 2016.

Oudeyer, Pierre-Yves, Baranes, Adrien, and Kaplan, Frédéric. "Intrinsically motivated learning of real-world sensorimotor skills with developmental constraints." In Gianluca Baldassarre and Marco Mirolli, eds., *Intrinsically motivated learning in natural and artificial systems*, 303–365. Springer, 2013.

Oudeyer, Pierre-Yves, Kaplan, F., and Hafner, V. V. "Intrinsic motivation systems for autonomous mental development." *IEEE Transactions on Evolutionary Computation*, 11(2), 265–286, 2007.

Pellis, Sergio, and Pellis, Vivien. *The playful brain: Venturing to the limits of neuroscience*. Oneworld, 2009.

Pfeifer, Rolf, and Bongard, Josh. *How the body shapes the way we think: A new view of intelligence*. MIT Press, 2006.

Piaget, J. *Play, dreams, and imitation in childhood*. Routledge, 1999.

Piek, J. P., and Carman, R. "Developmental profiles of spontaneous movements in infants." *Early Human Development*, 39(2), 109–126, 1994.

Pierson, Harry A., and Gashler, Michael S. "Deep learning in robotics: A review of recent research." *Computing Research Repository (CoRR), arXiv:1707.07217*, 2017.

Pinker, Steven. *The better angels of our nature: The decline of violence in history and its causes*. Penguin, 2011.

Pinker, Steven. *Enlightenment now: The case for reason, science, humanism, and progress*. Penguin, 2018.

Plummer, Robert Patrick. *A computer program which simulates sensorimotor learning*. Ph.D. thesis, the University of Texas at Austin, 1970.

Plummer, Robert Patrick. "A sensorimotor learning program." In *Proceedings of the 5th Australian Computer Conference*, 617–622. Australian Computer Society, 1972.

Pomerleau, Dean. "Neural network vision for robot driving." In M. H. Hebert, C. Thorpe, and A. Stentz, eds., *Intelligent unmanned ground vehicles*, 53–72. Springer, 1997.

Popper, Karl, Ryan, Alan, and Gombrich, E. H. *The open society and its enemies*. Princeton University Press, 2013.

Prince, C. G., Helder, N. A., and Hollich, G. J. "Ongoing emergence: A core concept in epigenetic robotics." In *Proceedings of the 5th International Workshop on Epigenetic Robotics*, Lund University Cognitive Studies, 63–70, 2005.

Purves, Dale, and Lichtman, Jeff W. "Elimination of synapses in the developing nervous system." *Science*, 210(4466), 153–157, 1980.

Quinlan, Philip T. *Connectionist models of development: Developmental processes in real and artificial neural networks*. Taylor & Francis, 2003.

Rizzolatti, Giacomo, Fadiga, Luciano, Gallese, Vittorio, and Fogassi, Leonardo. "Premotor cortex and the recognition of motor actions." *Cognitive Brain Research*, 3(2), 131–141, 1996.

Rochat, Philippe, and Striano, Tricia. "Perceived self in infancy." *Infant Behavior and Development*, 23(3–4), 513–530, 2000.

Rolf, Matthias, and Steil, Jochen J. "Goal babbling: A new concept for early sensorimotor exploration." In E. Ugur, Y. Nagai, E. Oztop, and M. Asada, eds., *Proceedings of IEEE International Conference on Humanoid Robots, Humanoids 2012 Workshop on Developmental Robotics*, 40–43, 2012.

Rosling, Hans, Rönnlund, Anna Rosling, and Rosling, Ola. *Factfulness: Ten reasons we're wrong about the world and why things are better than you think*. Flatiron Books, 2018.

Russell, Stuart, Dewey, Daniel, and Tegmark, Max. "Research priorities for robust and beneficial artificial intelligence." *AI Magazine*, 36(4), 105–114, 2015.

Russell, Stuart, and Norvig, Peter. *Artificial intelligence: A modern approach*. 3rd ed. Pearson, 2010.

Rutkowska, J. C. "Scaling up sensorimotor systems: Constraints from human infancy." *Adaptive Behaviour*, 2, 349–373, 1994.

Samuel, Arthur L. "Some studies in machine learning using the game of checkers." *IBM Journal of Research and Development*, 3(3), 210–229, 1959.

Samuel, Arthur L. "Some studies in machine learning using the game of checkers. II: Recent progress." *IBM Journal of Research and Development*, 11(6), 601–617, 1967.

Sandamirskaya, Yulia, and Storck, Tobias. "Neural-dynamic architecture for looking: Shift from visual to motor target representation for memory saccades." In *4th International Conference on Development and Learning and on Epigenetic Robotics*, 34–40. IEEE, 2014.

Schaeffer, Jonathan, Burch, Neil, Björnsson, Yngvi, et al. "Checkers is solved." *Science*, 317(5844), 1518–1522, 2007.

Schmidhuber, Jürgen. "Deep learning in neural networks: An overview." *Neural Networks*, 61, 85–117, 2015.

Shaw, Patricia, Law, James, and Lee, Mark. "Representations of body schemas for infant robot development." In *5th International Conference on Development and Learning and on Epigenetic Robotics*, 123–128. IEEE Press, 2015.

Shaw, Patricia H., Law, James A., and Lee, Mark H. "A comparison of learning strategies for biologically constrained development of gaze control on an iCub robot." *Autonomous Robots*, 37(1), 97–110, 2014; doi:10.1007/s10514-013-9378-4.

Sheldon, M. T. *Intrinsically motivated developmental learning of communication in robotic agents*. Ph.D thesis, Aberystwyth University, Wales, 2012.

Sheldon, M. T., and Lee, M. "PSchema: A developmental schema learning framework for embodied agents." In *Proceedings of the IEEE Joint International Conference on Development and Learning (ICDL) and Epigenetic Robotics*. IEEE, 2011.

Sherwood, Stephen L., ed. *The nature of psychology: A selection of papers, essays, and other writings by Kenneth J. W. Craik*. Cambridge University Press, 1966.

Shneiderman, Ben. *Leonardo's laptop: Human needs and the new computing technologies*. MIT Press, 2003.

Shultz, Thomas R. *Computational developmental psychology*. MIT Press, 2003.

Shultz, Thomas R. "A constructive neural-network approach to modeling psychological development." *Cognitive Development*, 27(4), 383–400, 2012.

Silver, David, Huang, Aja, Maddison, Chris J., et al. "Mastering the game of Go with deep neural networks and tree search." *Nature*, 529(7587), 484–489, 2016.

Silver, David, Hubert, Thomas, Schrittwieser, Julian, et al. "Mastering chess and shogi by self-play with a general reinforcement learning algorithm." *arXiv preprint arXiv:1712.01815*, 2017a.

Silver, David, Schrittwieser, Julian, Simonyan, Karen, et al. "Mastering the game of Go without human knowledge." *Nature*, 550(7676), 354, 2017b.

Smith, Linda, and Gasser, Michael. "The development of embodied cognition: Six lessons from babies." *Artificial Life* 11(1–2), 13–29, 2005.

Smith, Linda B., and Thelen, Esther. "Development as a dynamic system." *Trends in Cognitive Sciences*, 7(8), 343–348, 2003.

Smith, Peter K. *Children and play: Understanding children's worlds*, vol. 12. John Wiley & Sons, 2009.

Smith, Roger. "James Cox's silver swan: An eighteenth-century automaton in the Bowes Museum." *Artefact: Techniques, histoire et sciences humaines*, 4(4), 361–365, 2016.

Snyder, Lawrence H., Grieve, Kenneth L., Brotchie, Peter, and Andersen, Richard A. "Separate body-and world-referenced representations of visual space in parietal cortex." *Nature*, 394(6696), 887, 1998.

Spelke, Elizabeth S. "Principles of object perception." *Cognitive Science*, 14(1), 29–56, 1990.

Sporns, O., and Edelman, G. M. "Solving Bernstein's problem: A proposal for the development of coordinated movement by selection." *Child Development*, 64(4), 960–981, 1993.

Standing, Guy. *The corruption of capitalism: Why rentiers thrive and work does not pay*. Biteback Publishing, 2016.

Steels, Luc. *The Talking Heads Experiment: Origins of words and meanings*. Language Science Press, 2015.

Stork, David G. *HAL's legacy: 2001's computer as dream and reality*. MIT Press, 1998.

Sutherland, Stuart. *Irrationality: The enemy within*. Constable and Company, 1992.

Szegedy, Christian, Zaremba, Wojciech, Sutskever, Ilya, et al. "Intriguing properties of neural networks." *arXiv preprint arXiv:1312.6199*, 2013.

Tainter, Joseph. *The collapse of complex societies*. Cambridge University Press, 1988.

Tallis, Raymond. *The hand: A philosophical inquiry into human being*. Edinburgh University Press, 2003.

Tang, Yi-Yuan, Rothbart, Mary K., and Posner, Michael I. "Neural correlates of establishing, maintaining, and switching brain states." *Trends in Cognitive Sciences*, 16(6), 330–337, 2012.

Terada, Yuuzi, and Yamamoto, Ikuo. "An animatronic system including lifelike robotic fish." *Proceedings of the IEEE*, 92(11), 1814–1820, 2004.

Thaler, Richard H., and Sunstein, Cass R. *Nudge: Improving decisions about health, wealth, and happiness*. Yale University Press, 2008.

Thelen, E. "Grounded in the world: Developmental origins of the embodied mind." *Infancy*, 1(1), 3–28, 2000.

Thelen, Esther. "Rhythmical stereotypies in normal human infants." *Animal Behaviour*, 27, 699–715, 1979.

Tikhanoff, Vadim, Cangelosi, Angelo, and Metta, Giorgio. "Integration of speech and action in humanoid robots: iCub simulation experiments." *IEEE Transactions on Autonomous Mental Development*, 3(1), 17–29, 2011.

Tomasello, M., and Herrmann, E. "Ape and human cognition: What's the difference?" *Current Directions in Psychological Science*, 19(1), 3–8, 2010.

Topol, Eric J., and Hill, Dick. *The creative destruction of medicine: How the digital revolution will create better health care*. Basic Books, 2012.

Triantafyllou, Michael S., and Triantafyllou, George S. "An efficient swimming machine." *Scientific American*, 272(3), 64–71, 1995.

Tronick, E. "Stimulus control and the growth of the infant's effective visual field." *Perception and Psychophysics*, 11(5), 373–376, 1972.

Turing, A. M. "Computing machinery and intelligence." *Mind*, 59(236), 433–460, 1950.

Turing, Alan Mathison, and Copeland, B. Jack. *The essential Turing: Seminal writings in computing, logic, philosophy, artificial intelligence, and artificial life, plus the secrets of Enigma*. Oxford University Press, 2004.

Turkewitz, Gerald, and Kenny, Patricia A. "Limitations on input as a basis for neural organization and perceptual development: A preliminary theoretical statement." *Developmental psychobiology*, 15(4), 357–368, 1982.

Ugur, Emre, Nagai, Yukie, Sahin, Erol, and Oztop, Erhan. "Staged development of robot skills: Behavior formation, affordance learning, and imitation with motionese." *IEEE Transactions on Autonomous Mental Development*, 7(2), 119–139, 2015.

Vernon, David. "Enaction as a conceptual framework for developmental cognitive robotics." *Paladyn*, 1(2), 89–98, 2010.

Vernon, David, Metta, Giorgio, and Sandini, Giulio. "The iCub cognitive architecture: Interactive development in a humanoid robot." In *2007 IEEE 6th International Conference on Development and Learning*, 122–127. IEEE, 2007.

Vernon, David, von Hofsten, Claes, and Fadiga, Luciano. *A roadmap for cognitive development in humanoid robots*. Vol. 11 of *Cognitive Systems Monographs (COSMOS)*. Springer, 2010.

von Hofsten, C. "An action perspective on motor development." *Trends in Cognitive Sciences*, 8(6), 266–272, 2004.

Wainer, Joshua, Dautenhahn, Kerstin, Robins, Ben, and Amirabdollahian, Farshid. "Collaborating with Kaspar: Using an autonomous humanoid robot to foster cooperative dyadic play among children with autism." In *Humanoid Robots–10th IEEE-RAS International Conference*, 631–638. IEEE Robotics and Automation Society, 2010.

Walter, W. Grey. *The living brain*. Gerald Duckworth & Co, 1953.

Walter, W. Grey. "The brain as a machine." *Proceedings of the Royal Society of Medicine*, 50(10), 799–808, 1957.

Webb, William. *Our digital future: Smart analysis of smart technology*. CreateSpace Independent Publishing Platform, 2017.

Westermann, G., and Mareschal, D. "From parts to wholes: Mechanisms of development in infant visual object processing." *Infancy*, 5(2), 131–151, 2004.

Westheimer, Gerald. "Optical superresolution and visual hyperacuity." *Progress in Retinal and Eye Research*, 31(5), 467–480, 2012.

White, B. L., Castle, P., and Held, R. "Observations on the development of visually-directed reaching." *Child Development*, 35(2), 349–364, 1964.

Wiener, Norbert. *Cybernetics: or control and communication in the animal and the machine*. 2nd ed. MIT Press, 1961.

Wilkins, David. "Using chess knowledge to reduce search." In Peter W. Frey, ed., *Chess skill in man and machine*, 211–242. Springer, 1983.

Wilmott, Paul. "The use, misuse and abuse of mathematics in finance." *Philosophical Transactions of the Royal Society of London A: Mathematical, Physical and Engineering Sciences*, 358(1765), 63–73, 2000.

Wilmott, Paul. *Frequently asked questions in quantitative finance*. John Wiley & Sons, 2010.

Wolpert, Daniel M. "Probabilistic models in human sensorimotor control." *Human Movement Science*, 26(4), 511–524, 2007.

Woodward, Amanda L. "Infants' grasp of others' intentions." *Current Directions in Psychological Science*, 18(1), 53–57, 2009.

Yamada, Yasunori, Mori, Hiroki, and Kuniyoshi, Yasuo. "A Fetus and infant developmental scenario: Self-organization of goal-directed behaviors based on sensory constraints." In *Proceedings of 10th International Conference on Epigenetic Robotics. Lund University Cognitive Studies*, 149, 145–152. 2010.

Yosinski, Jason, Clune, Jeff, Nguyen, Anh, Fuchs, Thomas, and Lipson, Hod. "Understanding neural networks through deep visualization." *arXiv preprint arXiv: 1506.06579*, 2015.

Zeiler, Matthew D., and Fergus, Rob. "Visualizing and understanding convolutional networks." In *Computer Vision—European Conference on Computer Vision, 2014*, 818–833. Springer, 2014.

Zeki, Semir. *A vision of the brain*. Oxford University Press, 1993; now available at www.vislab.ucl.ac.uk.

Zuboff, Shoshana. *The age of surveillance capitalism: The fight for a human future at the new frontier of power*. Profile Books, 2019.

INDEX

Acquisitions
  AI companies, 271
  robotics companies, 271
Adams, Douglas, 19, 114
AI applications, 111, 161,
    274–277
Alan Turing Institute, 310
Algorithms, 17–18, 59–61, 66, 304,
    319n5
  developmental, 202–203, 205, 215,
    217
  financial, 288–289
  general AI (AGI), 151–153, 171
  learning, 100–101, 106, 164
Alibaba Group, 271
Amazon.com, Inc., 28–29, 38–39, 51,
    256, 261, 273
Apple Inc., 25, 38, 121, 261, 270,
    271, 310
Armstrong, Stuart, 296
Artificial General Intelligence (AGI),
    113, 124, 143, 257, 276
  predictions, 298–300
  for robots, 163, 170, 264

  start-up companies, 272–273
  theory, 151–156, 160–161
Artificial Human Intelligence (AHI),
    155–156, 161, 170–171, 262–264
Artificial Intelligence (AI)
  affective AI, 41–42, 243, 247
  definitions of, 7, 55–56
  human level (HLAI), 154–156,
    255–257
  limitations of, 11, 151, 298–299
  predictions, 16, 65–66, 162, 232, 268,
    293, 296
  symbolic and subsymbolic, 75
  task-based, 113, 156, 161–162, 170,
    197, 253–257, 262, 264
  weak and strong, 113
Artificial Neural networks (ANNs), 79,
    143–144, 155, 256
  auto-encoders, 138
  basics, 133–135
  Convolutional Neural Networks
    (CNNs), 88–91, 107–108, 259
  Generative Adversarial Networks
    (GANs), 139

Artificial Neural networks (ANNs) (cont.)
  integration, 158–160, 258
  issues with, 139–141, 161
  Recurrent Neural Networks (RNNs), 137, 159
Ashby, Ross, 172–173
Assessing robot functionality, 19–20
Associated sensory event, 235
Associative memory, 203, 217–220, 226
Automata, early robots, 4–5, 173
Automatic Teller Machine (ATM), 307
Automation
  agriculture, 23, 48, 50, 161, 276, 307
  job losses, 284, 306–308
  manufacturing, 46–47, 156, 165, 270, 277, 279, 284–285
Autonomous (driver-less) vehicles, 25, 275–276

Baidu Research, 272, 273, 310
Banavar, Guruduth, 285, 287
Bank of England, 306
Barlow, Horace, 172, 173, 204, 224
Bayes method, 105, 109, 153
Becker, Joseph D., 226
Berners-Lee, Sir Tim, 127
Bernstein, Nikolai, 196, 202
Big Data, 8, 15, 97–98, 102–104, 107, 109, 112, 127, 153, 158, 259, 261, 263–264
Biorobotics, robot fish, 40, 42
Black box, 140, 144, 162, 258, 328n17
Blue Brain Project, 128
Blue Gene (IBM), 126
Body models, 182, 240, 242–243, 311
Bosch Home Appliances, 21
Boston Dynamics, 44
Bostrom, Nick, 91, 157, 301
Brains
  and bodies, 9, 163
  cortex, 81–83
  as machines, 130–133
  models, 126–127, 171–173
  modularity of, 146
  neurons, 126
  simulation of, 128–130, 141–144
  states, 201
  structure, 125–126
  See also Neuroscience
Breakthrough effects, 19, 65, 86–88, 91–92, 106–109, 162
Bridle, James, 312
Brockman, John, 264, 296
Brooks, Rodney, 178
Bruner, Jerome, 179, 190

CERN, data handling, 127–128
Challenges and competitions, 53, 254–256, 264
  bin picking, 38–39, 254
  DARPA, 24, 51, 254
  RoboCup, 27, 51, 256
  visual recognition, 86–89
Chaos, in environment, 11, 20, 47–50, 53, 119, 144
Chatbots, 17, 117, 120, 256
Clark, Andy, 138, 178
Clarke, Arthur C., 325nn3–4 (chap. 8)
Cloud services, 30, 93, 108, 121, 258, 271, 277, 282–283
Cognition, 55, 75, 79, 153, 176–179, 190–191, 194, 240, 250, 313.
  See also Thinking
Cognitive Computing (IBM), 285
Cohen, Jack, 129
Collaboration and cooperation, 165, 250, 291, 294, 299
Common sense knowledge, 26, 72–73, 122–123
Complex system theory, 132, 144
Computational neuroscience, 134–136, 143–144, 183
Computer chess
  basic algorithm, 59–61
  compared with human play, 62–63

knowledge intensive, 63
tournaments, 64
world champion, 60–61
Computer vision, 83–86
object recognition competition, 86–89
Consciousness, 178, 244–246, 251, 298–299
Constraints
in AI, 58, 120
in developmental robots, 201–202, 205–206, 210, 212–214, 315
in engineering, 20, 32, 284
in infancy, 188–192
Constructivism, 133, 184, 206
Conversation, social, 8, 30, 121–124, 261–262, 311
Conversational systems, 6, 17, 52, 120, 250
Convolution filter, 83
Craik, Kenneth, 171–173, 180, 185
Crick, Francis, 245
Cybernetics, 171–175, 179
Cyc project, 72–74, 123, 263

Daimler Trucks North America LLC, 24
Damasio, Antonio, 163, 178
D'Andrea, Raffaello, 29
DARPA challenge, 24, 51, 254–256
Data mining, 97, 101–102, 109, 256
Dautenhahn, Kerstin, 9
Davis, Ernest, 249
Decision trees, 98–101
Deep Blue, chess computer, 60–61, 64–65, 106–107
Deep Learning, 19, 81, 89, 91–93, 104, 107, 257–258, 273–274
for robots, 259–260
DeepMind Technologies. *See* Google: DeepMind
DeepQA (IBM), 118

Deep Thought
chess program, 61
science fiction computer, 114
Degrees of freedom
of bodies, 195–196
of iCub, 208
de Latil, Pierre, 174
Deutsch, David, 128, 152, 336n5
Developmental psychology, 14, 182–183, 189, 232, 317
Developmental robotics, 13–14, 164, 181–182, 186, 198, 206–207, 251, 313, 316–317
Developmental timelines, 186–189, 201, 206, 210, 233, 315
de Waal, Frans, 294
DexNet, 259
Dialog systems, 32, 42, 48, 52, 120
Domain of discourse, 118–120, 124
Domestic robots, 22, 271
Domingos, Pedro, 152–153
Drescher, Gary L., 226
Dyson Ltd, 21

Edelman, Gerald, 245
Elsie and Elmer, 173
Embodied intelligence, 9, 178–179, 184
Embodiment, 12, 13, 123, 147, 163, 178–179, 264, 298–303, 311
Emergent properties, 132, 193, 250
Emotion AI, 42, 243. *See also* Artificial Intelligence: affective AI
Empathy, 42, 206, 243, 246, 298–299
Enaction, 169, 179–180, 184, 206, 303
End-to-end learning, 106–107, 109, 260
Engineering approach, 9–10, 31–33, 109, 113, 130–132, 140–141, 146, 278–281, 290
EPFL (École Polytechnique Fédérale de Lausanne), 126
Epigenetic robotics. *See* Developmental robotics

Ethical issues, 114, 141, 161, 259, 268, 276, 279–281, 286–290, 310
Evaluation function (evaluator), 58–61, 64, 68–69, 76–78, 90, 139
Experimental software, design of, 201–206
Expertise, 62–63, 66, 242, 285, 302, 336n9
Expert opinion, 280, 296–300

Facebook, Inc., 90
Fake news, 304
Farming and agriculture, 23, 276–277
Fauconnier, Gilles, 231, 249
Feature engineering, 77–79, 86, 92, 107
Feedback, 132, 137, 159, 171–172, 174–175, 227, 245
Fellgett, Peter B., 175
Fetal models, 187, 192–193
Field robotics, 22–23
Financial algorithms, 288–289
Frase, Peter, 308
Friendly AI, 310
Friendly robots, 4, 14, 51–52, 145, 164, 239
Future developmental robots, 237–241, 247–248

Gallese, Vittorio, 244
Game playing by computer
  AlphaGo, 90–91, 105–108
  AlphaGo Zero, 106–108
  AlphaZero, 106–109, 158
  Atari, 89, 106–107
  breakthroughs, 59, 61, 109
  checkers, 76–77
  chess, 58–66
  Deep Blue, 60–61, 64–65, 106–107
  DQN, (Atari), 89, 106–107
  Go, 90–91, 105–109
Gates, Bill, 37
Gazing and reaching, 13
Generative systems, 261–265

Gestures, 6, 14, 30, 42, 52–53, 164, 234–237, 246, 294
Goal directed behavior, 200–201, 233
Goal learning, 200, 235, 251
Gobet, Fernand, 62
Godfrey-Smith, Peter, 244
Good, Jack, 157, 172
GoodAI, 273
Google, 24–25, 38, 44, 103–104, 121, 258, 260, 271, 273
  DeepMind, 89–90, 105–109, 113, 139, 158, 286–287
  machine translation (GNMT), 274
GPU, 87, 108, 258
Grand, Steve, 130, 321n5
Grasping, 187–189, 209, 260. *See also* Hands
Grounding, situated, embedded, and embodied, 178
Growth, as methodology, 13–14, 191–192
Growth rates
  AI companies, 271–272
  exponential, 268–270
  robotics market, 270–271

HAL 9000, 10, 114–115, 157, 325n4 (chap. 8)
Haldane, Andrew G., 306
Hand-crafted knowledge, 88, 107
Hands, 53, 187, 189, 211, 259
  iCub, 208–210, 213–214, 236
Harari, Yuval Noah, 309
Harnad, Stevan, 117, 176
Hassabis, Demis, 108, 286
Hawking, Stephen W., 157
Hinton, Geoffrey, 86, 159
Hitchhiker's Guide to the Galaxy, The, 19, 114
Holland, Owen, 173, 330n5
Honda Motor Corporation, 36–37, 43–44, 52, 270
Hsu, Feng-hsiung, 61, 107

Hubel, David, 81–83
Hui, Fan, 90
Human and robot nature, 14–15, 282, 293
Human Brain Project (HBP), 126–128
Human Brain simulation. *See* Whole Brain Emulation (WBE)
Human characteristics, 162–163, 251, 294–295
Human-computer systems, 165, 284–285
Human Genome Project, 103
Human helpers, 45
Human level Artificial Intelligence (HLAI), 154–155, 255–257
Human-robot interaction, 42, 51, 53, 197, 277
Human world
  and digital world, 123, 295, 311
  facts of life and death, 162, 244–245, 251, 265, 296
Husqvarna Group, 20–21
Hutchins, Edwin, 75, 250
Hybrid AI systems, 148–149
Hyperacuity, 223
Hypotheses
  1 difference between AGI and AHI, 170
  2 importance of enactive, cognitive growth, 184
  3 play behavior, significance of, 200

IBM, 60–65, 77, 90, 118–119, 126, 271, 285
iCub robot, 208–209, 232, 236, 239, 259
Image processing, 7, 83–86, 92, 260, 277
Image recognition, breakthroughs, 86–89
Individuality, 238–239, 248

Infant development, 181–189, 199–202, 206, 216, 224–225, 235–236, 246, 248, 251
Infants, as inspiration, 13–14, 181, 206, 243, 317
Innovation, start-ups, 25, 31, 35–37, 53, 55, 264, 271–272
Integrated neural systems, 158–160
Integrity and transparency, 144, 165, 282, 287, 290–291, 304–306, 311–312
Intel, 25, 271
Intelligence
  cultural, 8, 112, 291
  definitions of, 55–56, 69
  embodied, 12, 163, 178
  explosion of, 155–157
  general purpose, 11–12, 113, 124, 151
  human, 154, 170, 179
  social, 8–9, 112, 163–164, 291, 311
  three kinds of, 7–9, 111–112
Internet of Things (IoT), 283–284
Inverse problems, 195
iRobot Corporation, 21

Johnson, Mark, 184
Joint attention, 4, 246–247
Josiah Macy Jr. Foundation, 174

Kahneman, Daniel, 69, 154
Kasparov, Garry, 60–66
Kindred Inc., 272
Kiva Systems, 29
Knowledge
  common sense, 72–74, 123
  kinds of, 70–71
  metadata, 72
  ontologies, 71–73, 79
  representation, 72, 149–150, 159–160
  symbolic and subsymbolic, 159–160
  used in search, 69–70, 79
Kozima, Hideki, 4, 319n1
Kubrick, Stanley, 114, 157

Kuenssberg, Laura, 303
Kuniyoshi, Yasuo, 193, 210, 227

LCAS algorithm, 205, 232
Learning
  algorithms, 100–101
  autonomous, 273–274
  as change, 95–96
  by classifying features, 77–78
  from data, 98
  decision trees, 98–101
  end-to-end, 106–109, 260
  in game playing, 76–77, 105–107
  nouns and verbs, 235–237
  reinforcement, 89, 106, 260
LeCun, Yann, 91, 159
Legg, Shane, 56, 151
LEGO robots, 27
Lenat, Douglas, 72–73
Levesque, Hector, 122
Levine, Sergey, 260
Limitations of Artificial Neural
  Networks, 139–141, 161–162
Linux, 288
Loebner prize, 116, 256
Long, John, 322n9
Longitudinal development, 215–217
Long-tailed data problem, 25
Loyalty cards, 97–98

Machine Intelligence Research Institute
  (MIRI), 157
Machine Learning (ML), 96–97,
  152–153, 325n2 (chap. 7)
Machine metaphor, 130–131
Machines versus living systems, 123,
  162, 244–245
Mackay, Donald, 172–173, 175
Mappings, as relations, 222–224, 319
Marcus, Gary, 130, 249
Mason, Paul, 307
Master algorithms, 152–153
Matrix, The, 142, 329n22

Maturana, Humberto, 175
Maturation, proximodistal, 191–193,
  196
McCarthy, John, 175
McCulloch, Warren S., 133, 173, 174
Medical robots, 26, 49, 140
Mental simulation, 244, 263–265
Mercedes-Benz, 23
Metta, Giorgio, 194, 208
Microsoft Corporation, 37–38, 90,
  261, 273
Minsky, Marvin, 175, 180
Mirror neurons, 234
Mobileye, 25
Models of self, 124, 164, 251
Modular systems and integration,
  146–149
Moorehead, Alan, 295
Mori, Masahiro, 323n27
Motivation, intrinsic, 197–198
Motor babbling, 198–200, 204–205,
  212, 216, 314

NASA, 141, 263
  space robots, Robnaut and
  Valkyre, 44
Natale, Lorenzo, 208–209
Neuroscience, 125–126, 134–136,
  182–184, 249, 317
  computational, 134–136,
  143–144, 183
Ng, Andrew, 310
Nori, Francesco, 208
Norvig, Peter, 75
Novelty, 198, 202–205, 233, 315
Nudge theory, 306
Numenta, 272, 273
nuTonomy, 25

Object perception, 215, 225
  proto-objects, 222, 224–225, 233, 249
O'Neil, Cathy, 289
Ontogeny, 181, 184–185

Ontology, 71–73, 79
OpenAI Inc., 273
OpenCog, 161
Open source, 27, 73, 209, 258, 287–288
Orwell, George, 305
Oudeyer, Pierre-Yves, 198, 250

Panasonic Corporation, 275
Pask, Gordon, 171
Pattern discovery, 101–102
Pfeifer, Rolf, 178
Philips robot dog, 174
Physical symbol system hypothesis, 176–177
Piaget, Jean, 186, 199, 226
Pinker, Steven, 264, 309
Pitts, Walter H., 133
Play, 198–201
  algorithm, 202
  behavior, 217–221
  definition, 199
  importance of, 198, 249
Plummer, Robert P., 226
Pointing and gesture, 246, 301, 316
Popper, Karl, 152, 329n10
Predictive coding, 138
Probability, 104–105
Product evaluation, 281–282
Pronoun problems, 122
Proprioception, 209, 211, 218, 233
Proto-objects, 215, 225

Ratio Club, 172–173
Reaching, 188–189, 217, 233–234, 301
Rees, Martin, Lord Rees of Ludlow, 157
Regulation, 161, 279–280, 284, 290–291, 305–306, 310
Representations
  growing, 184, 187, 193
  integrated, 159–160
  knowledge, 72, 95, 148–150
  learning, 159, 256
Responsibility, 280, 282, 290, 298–299

Rethink Robotics, 45
Reward, 89, 106, 197, 201, 316
Robotics
  assistive, 41
  general-purpose, 50, 52, 151, 163–164
  humanoid, 36, 42–45, 193–195
  industrial, 46–48, 254
  research and development, 35–38
  service, 22, 49, 256, 271
  simulation in, 227
  social, 6, 52, 256
  swarm, 27–28
  task difficulty, 49–50
Robots, companion, 30–31
  Kirobo Mini, 30, 32, 52
  MiRo, 31
  Paro, 31
Robots, domestic, 22, 254–256, 270–271
Robots, entertainment, 29–30
  AIBO, 29–30, 250
  Cozmo, 30, 32, 52
  NAO, 44
  soccer, 27, 29, 43, 255–256
  Vector, 30
Robots, humanoid, 42–45
  Honda ASIMO, 36, 43, 52
  Honda E2-DR, 44
  Honda P3, 36, 52
  iCub, 13-14, 208–209
  Keepon, 3–4, 6, 30, 52
  Kilobot, 27–28
  Kismet, 9
  Sony QRIO, 37
Robots, industrial
  Baxter, 45
  cobots, 277
  SecondHands, 45–46
Robots, medical
  da Vinci surgical robot, 26
Rosling, Hans, 309
Russell, Stuart, 75, 287
Rutkowska, Julie, 190

Samuel, Arthur, 76–77, 96
Sandini, Giulio, 194, 208
Scaffolding, 191, 237–238, 316
Scaremongering, 15–16
Schema memory, 226–227
Schmidhuber, Jürgen, 137
Scientific reductionism, 131–132, 144, 153, 240
Searching and planning
  applications, 74–75
  basic technique, 56–58
  Monte-Carlo tree search, 105
  trade-off with knowledge, 67–69
  using knowledge, 69–70
Se-dol, Lee, 90
Self-awareness, 186, 241–246, 248, 251, 294, 298
Self-driving vehicles, 23–26, 274–276
Self models, 15, 124, 164, 239–243, 251
Semantic Web, 72–73
Sense of self, 123, 163, 207, 295
Sensorimotor maturity, 211–213
Sentience, 244, 251, 303
Shaw, George Bernard, 303
Sheldon, Michael, T., 227
Shneiderman, Ben, 312
Shultz, Thomas, 317, 328n15
Singularity, the, 156–157, 296, 300–301
Social Engineering, 304–306
Social interaction, 6–9, 13–14, 52–54, 112, 121–124, 163–164, 246, 311
Social robots, 6, 9, 30, 33, 42, 52–54, 123, 161–164, 212, 256, 260–262, 264–265
Softbank Robotics, 44
Soft computing, 176, 240, 242
Software, definitions, 17–18
Software engineering, 112–113, 124, 269
Sony Corporation, 29–30, 37, 250
Space
  All centric, 236, 248
  Egocentric, 193, 222, 233, 236, 248

Gaze, 213–217, 221–224
Reach, 194, 214–217, 220
Selke, Elizabeth, 225
Standards, 278–281, 284, 286–291, 305, 311–312
Standing, Guy, 307
Steels, Luc, 250, 274
Stewart, Ian, 129
Stockfish, 64, 106, 108
Subliminal perception, 304–305
Suleiman, Mustafa, 287
Superintelligence, 155–158, 296, 299–301, 310
Super recognizers, 257
Suspicious coincidences, 204, 224–225
Sutton, Richard, 141, 272
Symbol grounding, 176–177, 163, 182, 184

Tainter, Joseph, 308
Tallis, Raymond, 259
Task-based methodology, 113, 253–254
  limitations, 151, 160–162, 170
Teaching robots, 227, 237–239, 251
Tenenbaum, Joshua, 249
TensorFlow, 108, 258
Tesco plc, 97–98, 325n3 (chap. 7)
Tesla, Inc., 25, 275
Thelen, Esther, 196, 227, 313
Thinking, 55
  fast and slow, 69, 154
  human and machine, 62–63, 154
  in the wild, 250
Threats and dangers, 291, 300–310
  dystopia, 308–309
  loss of jobs, 306–308
  non-AI, 310
  social engineering, 304–306
  superintelligence, 300–302
  transhumanism, 302–303
Timelines
  developmental, 186–189
  for iCub, 210–211

INDEX

Toyota Industries, 25, 30–31, 44, 147
TPU (Tensor Processing Unit), 108, 258
Training robots, 227, 237–238, 251
Transhumanism, 298, 302–303
Transparency, 282, 287–290, 305–306
Trust, 278–284, 290–291
Turing, Alan M., 172, 157
  on chess, 59, 65
  on "growing" AI, 185
  on thinking, 56, 118
Turing test, 115–120, 262
Tversky, Amos, 324n1 (chap. 5)
2001, A Space Odyssey, 10, 114, 157, 325n5 (chap. 8)

Ultraintelligent machine, 157
Uncertainties, 20, 32, 47–50, 104, 109, 119, 282, 296
Unstructured environments, 38, 47, 197

Validation and Verification, 75, 140–141, 144, 258
Varela, Francisco, 175
Vernon, David, 169, 179, 185, 194, 208
Vicarious Inc., 273
Virtual assistants, 31, 261
Visual processing
  in the brain, 81–83
  breakthrough, 86–89
  by computer, 83–86, 92
Von Hofsten, Claes, 199

Walter, W. Grey, 172–173
Warehouse robots, 28, 48–49, 277
Watson (IBM), 7, 10, 118–120, 147, 285–286
Waymo LLC, 24
Whole Brain Emulation (WBE), 128–130, 144, 301
  issues, 129–130
Wiener, Norbert, 171–173
Wiesel, Torsten, 81–83

Wilmott, Paul, 289
Winograd schemas, 122

Yamaha Corporation, 147
YouTube, 103

Zuboff, Shoshana, 305